Essential Immunology

Essential Immunology

Jaxson Peterson

MURPHY & MOORE
www.murphy-moorepublishing.com

Published by Murphy & Moore Publishing,
1 Rockefeller Plaza,
New York City, NY 10020, USA
www.murphy-moorepublishing.com

Essential Immunology
Jaxson Peterson

International Standard Book Number: 978-1-63987-202-2 (Hardback)

Cataloging-in-Publication Data

Essential immunology / Jaxson Peterson.
p. cm.
Includes bibliographical references and index.
ISBN 978-1-63987-202-2
1. Immunology. 2. Serology. I. Peterson, Jaxson.
QR181 .E87 2022
616.079--dc23

TABLE OF CONTENTS

PREFACE

The branch of biology which focuses on the immune system of various organisms is termed as immunology. It involves the process of charting, measuring and contextualizing the physiological functioning of the immune system in states of both health and disease. The chemical, physical, and physiological characteristics of the components of the immune system are also studied under it. The discipline of immunology can be broadly categorized into classical immunology, clinical immunology and developmental immunology. It finds applications in numerous branches of medicine, especially in the domains of organ transplantation, oncology, rheumatology, virology, bacteriology and psychiatry. Immunotherapy is defined as the use of antigens and other immune system components to treat a disease or genetic disorder. In order to achieve the desired therapeutic goal, modern immunology has become increasingly selective in its manipulation of the body's immune system. This book is a valuable compilation of topics, ranging from the basic to the most complex theories and principles in the field of immunology. While understanding the long-term perspectives of the topics, it makes an effort in highlighting their impact as a modern tool for the growth of the discipline. This book attempts to assist those with a goal of delving into the field of immunology.

A detailed account of the significant topics covered in this book is provided below:

Chapter 1- Immunology can be defined as the branch of biology concerned with the structure and function of the immune system which is responsible for the defense of the body against infections. This ability of the body to defend itself against infections is called immunity. Immunology and its related concepts like immunity, types of immunity, immunization and its types are the topics introduced to the reader in this chapter.

Chapter 2- Antigens are the substances which evoke an immune response in the body. Antibodies are specialized proteins produced by a body's immune system in response to the introduction of an antigen. There are different types of antigens and antibodies. This chapter is dedicated to the close study and examination of the subject matter for the reader's benefit.

Chapter 3- A T-Cell is a type of white blood cell produced in the bone marrow and maturing in the Thymus. A B-Cell is also a type of white blood cell which matures in the bone marrow. Their difference lies in their reaction to the introduction of an antigen. T-Cells trigger the B-Cells to develop plasma cells and activate T killer cells that are ultimately responsible for killing and removing the pathogen. This chapter has been carefully written to provide a better perspective of T and B lymphocyte cells.

Chapter 4- Myeloid cells are blood cells that are produced in the bone marrow. They are activated upon the introduction of an antigen and play a major role in innate immunity. The complement system is a complex system of soluble proteins, enzymes and receptors that act in function together to help eliminate infectious microorganisms. Both the myeloid and complement system play a vital role in the immune system and are comprehensively covered in this chapter.

Chapter 5- Cytokines are a large group of proteins, peptides and glycoproteins secreted by the cells of the immune system. They play the role of a signaling molecule, synchronizing the responses of the immune system. Cytokine nomenclature, cytokine receptor, cytokine regulation, signal transduction and pathways, interferon, etc. are a few other terms which are discussed in detail in this chapter.

Chapter 6- Like any other functional system of the body, the immune system is prone to disorders which affect its functions and can sometimes be fatal. Some disorders which may plague the immune system are autoimmune diseases, hypersensitivity, immunodeficiency, transplant rejection, autoimmunity, HIV/AIDS, etc. This chapter aims to provide an in-depth understanding of these immune system disorders.

It gives me an immense pleasure to thank our entire team for their efforts. Finally in the end, I would like to thank my family and colleagues who have been a great source of inspiration and support.

Jaxson Peterson

Chapter 1

Immunology: An Introduction

Immunology can be defined as the branch of biology concerned with the structure and function of the immune system which is responsible for the defense of the body against infections. This ability of the body to defend itself against infections is called immunity. Immunology and its related concepts like immunity, types of immunity, immunization and its types are the topics introduced to the reader in this chapter.

Immunology is the study of the immune system of higher organisms in relation to disease. Specifically, immunology can be defined as the branch of biology that is concerned with the structure and function of the immune system, the bodily distinction of self from non-self, and the use of antibody-based laboratory techniques or immunoassays. In general, the immune system of higher organisms can be broken down into two primary response systems that work together to create immunity. The two primary response systems are innate and adaptive immune responses, with the latter further divided into cell-mediated and antibody-mediated responses. The cell-mediated response is produced when a subset of sensitized white blood cells or lymphocytes directly attack material (e.g., usually a cell or a virus) that has been determined to be foreign to the body. The antibody-mediated response involves the transformation of a subset of lymphocytes into cells that produce and secrete specific antibodies against the foreign material. These two immune responses are triggered when foreign material is introduced into the host as depicted in figure.

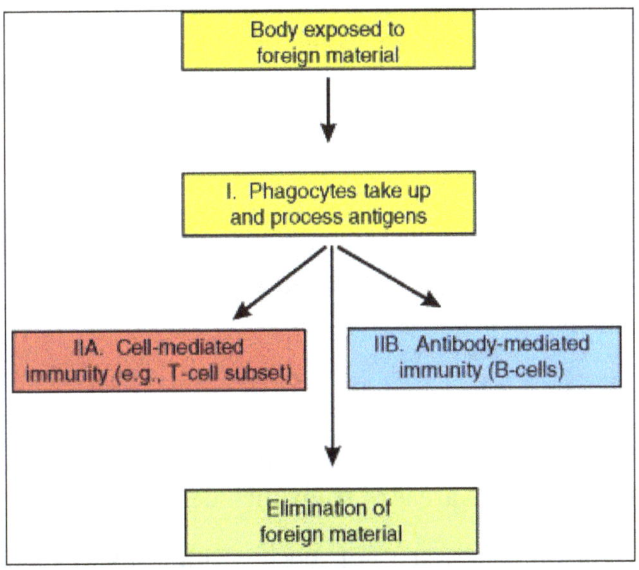

Flowchart showing the two major immune response systems that make up the host response to foreign materials. These two branches of the immune system work together to create immunity. The cell-mediated response is designed to directly attack and destroy cells determined by the body to be non-self. The antibody-mediated response is the branch of the immune system involved in the formation of antibodies.

Classical Immunology

Classical immunology deals with the fields of epidemiology and medicine. Immunology alone is the study of response of an organism to antigenic challenge and its recognition of what is self and what is not. It deals with the defence mechanisms including all physical, chemical and biological properties of the organism that help it to combat its susceptibility to foreign organisms, material, etc. Classical immunology deals with the relationship between the body systems, pathogens, and immunity.

Studying the Immune System

The immune system is studied as per its molecular and cellular components. Their function and interaction form the core of study of immunology as we know today. The immune system is divided into those which are static, or innate to the organism, and those which are responsive, or adaptive to a potential pathogen or foreign substance.

To maintain the integrity of our organism, it is essential to distinguish between self and non-self. Early in evolution, simple multicellular organisms developed a defence system activated by sensing typical molecules and chemical structures associated with pathogens. This system is evolved to its present day immunology. This innate, prefabricated, one-size-fits-all immune system is immediately available. It stops the pathogens and keeps an infection in check for a few days. After the first defence is overwhelmed the adaptive immune system kicks in.

The humoral response is a response of antibodies produced by the body to antigens. Antibodies are specific proteins released from a certain class of immune cells (B lymphocytes). Antigens are proteins or chemicals or even living organisms that elicit generations of antibodies, hence they are called Antibody Generators. Immunology is the study of these two biological entities.

Once exposed to antigens the body mounts a cellular response. This can kill infected cells in its own right and also controls the level of antibody response. The whole system is a highly balanced one and there are disorders of both under-active as well as over-active immune system. Both activities are highly interdependent.

In this century immunology has spread its frontiers with much research being performed in the more specialized niches of immunology. This includes the immunological function of cells, organs and systems not normally associated with the immune system along with studies of classical models of immunity.

Evolutionary Immunology

Evolutionary immunology studies the evolution of immunity and the immune system over ages among animals and plants. The main areas of study include:

- The study of relationships between nonspecific (innate) and specific (acquired, or adaptive) immunity.
- The study of how graft rejection occurs.
- The study of occurrence and historical development of T- and B-systems of immunity.

Evolutionary immunology also looks at different cell types of the immune system and the historical development of the ability to recognize foreign antigenic material.

It is known that life on Earth in the beginning was in the form of single-cell organisms more than 3.5 billion years ago. With development of multicellular organisms the continuous interactions between the diverse species was both in terms of cooperation and competition. The struggle to survive led to the evolutionary refinement of intricate immune systems. This included the development of both the adaptive and innate immune system.

Evolution of the Immune System

Evolutionary immunology clues have been obtained from studies of humans and mouse models. The core concepts of innate immunity for example have been noted across species of animals.

The observations of phagocytosis by mesenchymal cells in starfish and sea urchins led to the original conceptualization of self-versus non-self-recognition.

Birds provided clues to cellular and humoral immune function to the separate T and B lineages of lymphocytes and also led to the discovery of genes and their specific receptors involved in adaptive immunity. It was experiments on birds that gave clues to gene conversion and diversification of antibodies.

Now the concept of Toll receptors as crucial mediators of innate immunity has been developed from studies in insects. These animal studies provide clues to evolutionary immunology as well.

Innate and Adaptive Immunity

In the history of development of immunity it has been seen that innate immunity preceded adaptive immunity in the evolution of immune recognition and it is the common link between immune recognition in invertebrates and vertebrates.

It is in vertebrates that different clones of lymphocytes against specific antigens are seen. The vertebrates also have antigen-specific receptors of two types. Those in jawed vertebrates, sharks to humans, are based on the rearrangement of segmental elements encoding the immunoglobulin domains in T and B cell receptors mediated by recombinase-activating proteins RAG-1 and RAG-2. Jawless vertebrates such as lamprey and hagfish have a system that works in a RAG-independent way.

Among invertebrates immune-type receptors may undergo somatic differentiation through RNA processing mechanisms. This helps them achieve adaptive protective immunity and not true immunological memory. Polymorphic diversification for self-non-self-recognition is seen even in marine invertebrates.

Developmental Immunology

The immune system develops from hematopoietic precursor cells in the bone marrow. These precursors go through rounds of cell division and differentiation to give rise to the various different lymphoid and non-lymphoid lineages that go to make up the innate and adaptive immune systems. Myeloid cells of the innate immune system develop under the control of a variety of hematopoietic growth and differentiation factors which are orchestrated to result in the continuous production of the appropriate amounts of each cell type. Lymphoid cells of the adaptive immune system also start their development in the bone marrow. B lymphocytes responsible for the eventual production of antibodies develop along a pathway that requires rearrangement of the immunoglobulin genes to generate the vast array of individual B cell specificities and indeed, the normal progression of B cell development is regulated by the expression of immunoglobulin molecules. The other major component of the adaptive immune system, the T lymphocyte, develops in the thymus from precursor cells produced in the bone marrow. T cell development also requires rearrangement of genes that encode antigen specificity and selective events in the thymus ensure that mature T cells normally have the capacity to recognize foreign pathogen without recognizing self-antigens.

Clinical Immunology

Immunology pervades the activities of most clinicians. This is scarcely surprising since so much human disease results from the failure to combat infectious agents or from allergic reactions to antigens in the form of drugs, chemicals, or simple flora and fauna. The importance of immunological mechanisms in the pathogenesis of disease is reflected in the context of prophylactic immunisation or the therapeutic manipulation of the immune response. These intellectual and practical considerations have provided the stimulus for establishing the new discipline of clinical immunology.

Clinical immunology is thus the study of:

- Diseases caused by disorders of the immune system – this could be due to failure (under-activity), abnormal activation or malignant or cancerous growth of the cellular elements of the system.
- Pathology of other disorders where immune reactions play a part.
- Medications and drugs that modify or modulate the immune system for example those used in transplant rejections.
- Vaccines and other agents that modify immune reaction to specific pathogens.

Disorders of the Immune System

The diseases caused by disorders of the immune system fall into broad categories:

- Immunodeficiency – This occurs if the immune system fails to provide an adequate response to pathogens. This may be seen as a response to a long term disease condition like diabetes, chronic granulomatous disease or due to an infection such as HIV AIDS (Human Immunodeficiency virus causing Acquired Immunodeficiency syndrome).
- Autoimmunity – This occurs when the immune system attacks its own host's body. Some examples include systemic lupus erythematosus, rheumatoid arthritis, Hashimoto's disease and myasthenia gravis.
- Hypersensitivities and allergic reactions are also a part of immune system disorders. Here

the system responds inappropriately to harmless compounds for example in asthma and allergies or may respond too aggressively for example in anaphylaxis reactions.

Veterinary Immunology

Veterinary immunology is a branch of Immunology dedicated to improving animal health. Like humans, animals also suffer from diseases caused either when organisms try to invade their body, or when their immune system does not function properly. Wild, domestic, and farm animals are commonly exposed to a whole range of dangerous bacteria, viruses and parasites, which threaten their welfare. Animal infections can have widespread effects on human working sectors, like food and agriculture. Moreover, many animal infections can be naturally transmitted across the species barrier to infect humans and vice-versa, a process termed zoonosis. For example, well-studied infections including swine and avian influenza, as well as, malaria and Lyme disease are due to transmission from animals and insects to humans. It is therefore extremely important that these types of diseases are effectively controlled. These measures not only prevent any further transmission to other animals and humans, but also reduce any potentially devastating social and economic consequences.

Neuroimmunology

Neuroimmunology encompasses fundamental and applied biology, immunology, chemistry, neurology, pathology, psychiatry and virology of the central nervous system (CNS). Scientists in the field study the interactions of the immune and nervous system during development, homeostasis and response to injuries with the major aim of developing approaches to treat or prevent neuroimmunological diseases.

The immune system has been generally regarded as autonomous and the brain protected by the blood–brain barrier, (BBB) and in the words of Rudyard Kipling, 'never the twain shall meet'. In the past decades these dogmas have been strongly challenged and dispelled with the wealth of evidence showing that not only does the nervous system receive messages from the immune system, but that signals from the brain regulate immune functions that subsequently control inflammation in other tissues. Communication between the immune system and the CNS is exemplified by the finding that many molecules associated with the immune system are widely expressed and functional in the nervous system and vice versa. Cross-talk between microglia and neurones is known to be essential for maintaining homeostasis, yet such cross-talk also occurs between oligodendrocytes and microglia. Disturbance in this communication due to peripheral infections in mice are known to trigger microglia activation and augment neurodegeneration. Similarly, recent experimental studies show that maternal infections lead to long-term changes in microglia and abnormal brain development in the offspring.

While current therapies aim to modulate neuroinflammation arising during the disease, future approaches should aim at disease prevention. For some diseases, the aetiological agents are known, and thus vaccination strategies are key for disease prevention. In other cases, the specific genes or environmental agents triggering disease require clarification. Prophylactic approaches for genetic disorders could exploit genetic modification during development, while cell therapy strategies may aid regeneration of the damaged nervous system. Exploitation of infections agents may also be beneficial, as demonstrated by the recent clinical trial using a non-pathogenic poliovirus for treating glioblastomas.

For disease prevention, rapid and specific diagnoses as well as adequate ways to monitor the disease course and response to therapy are crucial. Thus, advancements in biomarker research will be key to faster diagnosis and more efficient monitoring in clinical trials, speeding up drug development and reducing costs. Biomarkers of neuroimmunological diseases may include markers of BBB disruption, demyelination, oxidative stress and excitotoxicity, axonal/neuronal damage, gliosis, remyelination and repair, but should also focus on markers of altered immune function such as cytokines, chemokines, antibodies, adhesion molecules, antigen presentation and changes in cellular subpopulations. Ideally, detection and collection of new biomarkers will be minimally invasive, specific for the disease and reflect response to therapy. Additionally, well-characterized tissue biobanks will be crucial to these advancements in biomarkers.

Another important aspect of future neuroimmunology research and developments will be in disease modelling. The highly effective CRISPR/cas9 system will allow precision engineering of the genome, and has the potential to speed up the generation of transgenic animal models, generating single-gene mutations in adult animals. Model systems making use of iPSCs from patients will also allow better translation of fundamental research data to the clinic. Further advances in CRISPR/cas9 or similar systems to increase transgene efficiency or to regulate gene expression using inducible expression systems will allow genes to be regulated once gene editing is completed. Such approaches will herald better treatments in the form of personalized medicine, gene editing (taking into account the ethical issues) and improved clinical trial design.

The increase in data generated by next-generation sequencing is expected to aid identification of genetic variants in neuroimmunological diseases. Such data are already contributing to designing algorithms, development of pharmacogenomics and personalized medicine. These approaches will be fundamental in reducing risks in drug development by avoiding adverse drug reactions, and minimizing cost by limiting drug administration solely to those patients who will benefit. While drug discovery is increasingly costly and prolonged, artificial intelligence (AI) may be key to reversing this trend. AI will use previously collected data and molecular dynamic predictions to reduce the number of compounds to be screened, repurpose compounds, predict interactions between compounds and their target and refine clinical trial populations. Advancements in targeted drug delivery will also reduce side-effect profiles of compounds and aid in those compounds that will readily cross the BBB. Both Big Pharma and academia have the potential to increase drug discovery efficiency by embracing AI, pharmacogenomics, personalized medicine and targeted drug delivery to provide future treatments of neuroimmunological diseases.

Immunity

Immunity to a disease is achieved through the presence of antibodies to that disease in a person's system. Antibodies are proteins produced by the body to neutralize or destroy toxins or disease-carrying organisms. Antibodies are disease-specific. For example, measles antibody will protect a person who is exposed to measles disease, but will have no effect if he or she is exposed to mumps.

Immunity can be described as either active or passive, depending on how it is acquired:

- Active immunity involves the production of antibodies by the body itself and the subsequent development of memory cells.
- Passive immunity results from the acquisition of antibodies from another source and hence memory cells are not developed.

Active immunity will result in long-term immunity but passive immunity will not (due to the presence or absence of memory cells).

- Both active and passive immunity can be induced by either natural or artificial mechanisms.

Examples of Active Immunity

- Natural – Producing antibodies in response to exposure to a pathogenic infection (i.e. challenge and response).
- Artificial – Producing antibodies in response to the controlled exposure to an attenuated pathogen (i.e. vaccination).

Examples of Passive Immunity

- Natural – Receiving antibodies from another organism (e.g. to the foetus via the colostrum or a new-born via breast milk).
- Artificial – Receiving manufactured antibodies via external delivery (e.g. blood transfusions of monoclonal antibodies).

Types of Immunity

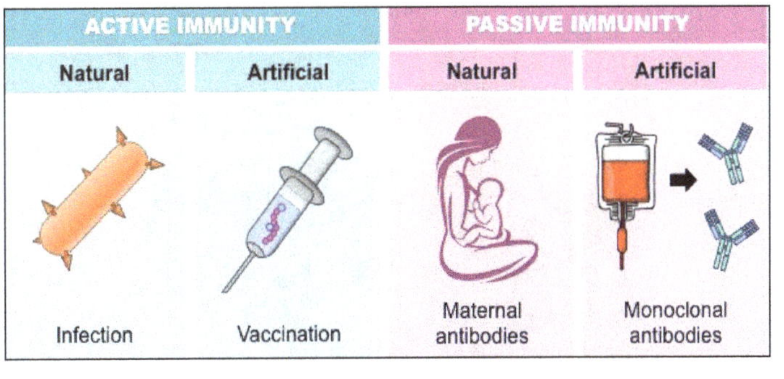

Innate and Acquired Immunity

There are two types of Immunity:

- Innate,
- Acquired.

Innate immunity and Acquired immunity complement each other. The response of the innate immunity system is fast. It has some specificity for microbes. It has no memory. Every time the innate immunity system encounters an antigen, it reacts ab initio. Hence, the innate immune response is not improved by repetitive invasion by the same antigen. The above three features distinguishes it from acquired immunity. Acquired immunity is also called Adaptive immunity and it consists of antibody mediated and T-cell mediated responses.

Both Soluble Factors and some Cells Participate in the Innate Immune Response

Innate immune system is present right from the birth. So, before acquired immune system develops, it protects the infant. In fact, this system changes very little during the lifetime. Inflammatory response to an infection is also part of the innate immune response.

The cells responsible for the innate immune response are:

- Phagocytes,
- Natural Killer Cells (NKC),
- Mast Cells,
- Dendritic Cells.

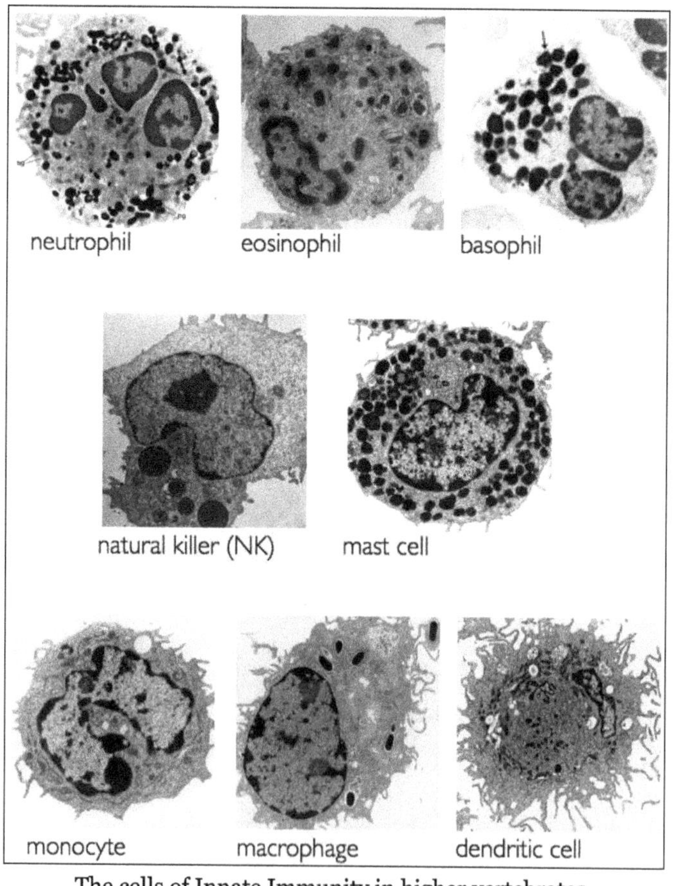

The cells of Innate Immunity in higher vertebrates.

- Natural Killer Cell (NKC).

- The NKC (lower left) is locating, binding and attacking a large cancer cell.

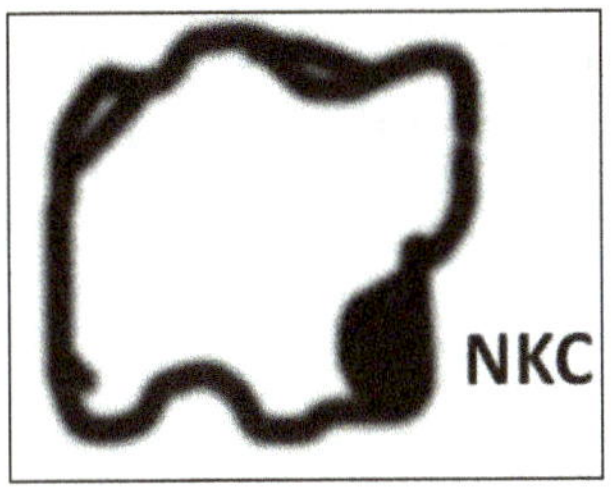

- The NKC has punctured and destroyed a large tumor cell.

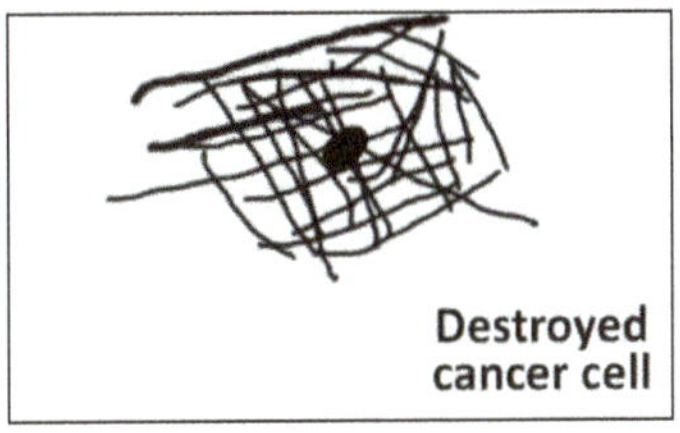

Molecules responsible for the innate immune response. Several classes of molecules also participate in the innate immune response:

Antimicrobial Peptides

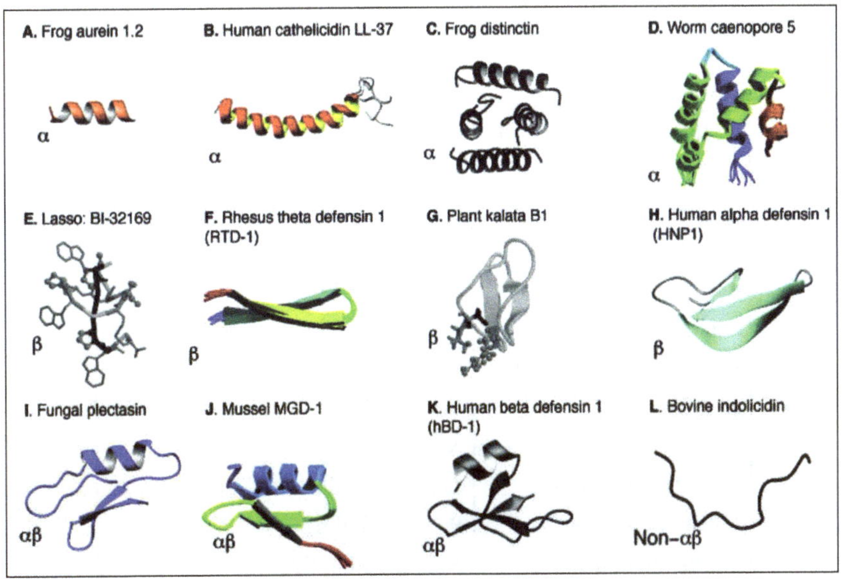

Examples of Antimicrobial Peptides.

Epithelial cells of the respiratory, gastrointestinal and genitourinary tracts produce these antimicrobial peptides. If the microbial infection breaches the barrier of the epithelial cells, it encounters phagocytic cells One of the modus operandi of phagocytic cells during innate immune response is to produce some antimicrobial peptides. These peptides are made up of less than 60 amino acids They may have a α-helical structure, a β-sheeted structure, a αβ-mixed structure or just a nonαβ extended structure.

Properties of Antimicrobial Peptides (AMPs)

AMPs cover a wide spectrum of antimicrobial properties. This range includes gram positive and gram negative bacteria, virus, fungi. Many have net positive charge. Mechanism of action includes altering the membrane permeability of the cells of the pathogen. Some AMPs exploit chemotaxis and attract T-cells (part of the acquired immune system) and dendritic cells. Thus AMPs are part of the co-operation between innate and acquired immune system.

Illustrative List of AMPs

AMP	Secreted by
Defensins	
α-defensin	Mostly by neutrophils but some epithelial cells also secret
β-defensin	Mostly epithelial cells and keratinocytes
Cathelicidins	Macrophages, PMNs, also some epithelial cells
Psoriasin	Keratinocytes
Calprotection	PMNs

AMP in keratinocytes	AMP first ID as protease inhibitors
Cathelicidins	Cathelin
β-defensis	SLPI
BPI	SKALP
Lactoferrin	P – cystatin α
Lysozyme	Cystatin C
Dermcidin	
RNase 7	
AMPs in infiltrating cells	**AMP first ID as chemokines**
Cathelicidins	Psoriasin
α-defensis	CXCL9
granulysin	IP 10 (CXCL 10)
perforin	I-TAC (CXCL 10)
eosinophil cationic	**AMP first ID as nueropepetides**
protein (ECP)RNase 3	α- MSH
RANTES	Bradykinin
CTAP- 3	Neurotensin
Platelet basic protein	Chromogranin A,B
Fibrinopeptide A, B	Proenkephalin A
Thymosin b- 4	Neuropeptide YY
	Adernomeddullin

It is believed that we may not have yet discovered all AMPs. AMPs are either produced constitutively or produced via pathogen associated molecular patterns. The latter operates via pattern recognition receptors which are present on the following cells of the innate immune system. Note the relationship between cells and soluble factors of the innate immune system.

- Macrophages,
- PMNs,
- Myeloid and plasmacytoid dendritic cells,
- Some epithelial cells.

This mechanism is responsible for the limited specificity of the innate immune system. It results in removal of microbes with common or similar surface structural motifs through a single functional response. Apart from this, some soluble pattern recognition receptors also exist. All those receptors can also bind to the self-components and participate in apoptosis (clearing dead cells).

Various Pattern Recognition Receptors (PRRs)

Toll-Like Receptors (TLRs)

These are among the ancient receptors from the evolutionary perspective Initially Toll receptor was discovered in the fruit fly Drosophila and later various Toll-like receptors were found in the mammalian systems. While most of the TLRs are present on the cell surface (especially on mono nuclear phagocytes), some are also present on endosomes.

The family of TLRs recognize wide range of pathogens. TLRs are integral proteins having intracellular domains. Upon binding to the ligands on the pathogen, the cells containing TLRs produce cytokines. Thus the consequences are:

- Inflammatory response due to cytokines which include TNF-α and IL-2.
- Enhances the capacity of the cells to kill microbes.
- Augments its antigen presentation capacity.

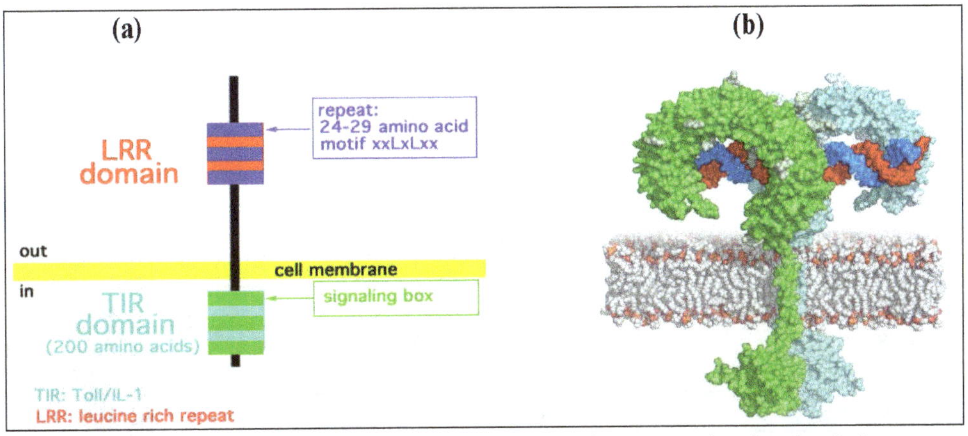

Structure of a Toll like Receptor (a) Schematically (b) using molecular simulation.

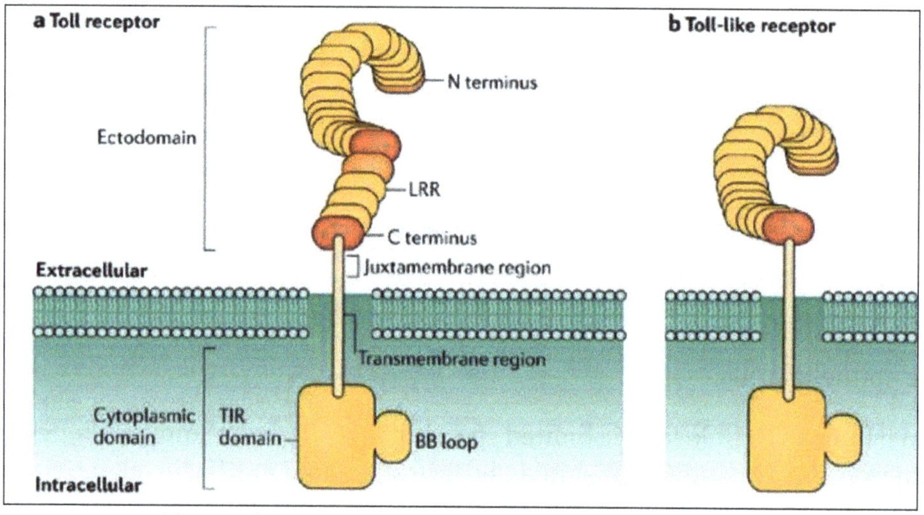

Difference between (a) Toll receptor and (b) Toll like receptor.

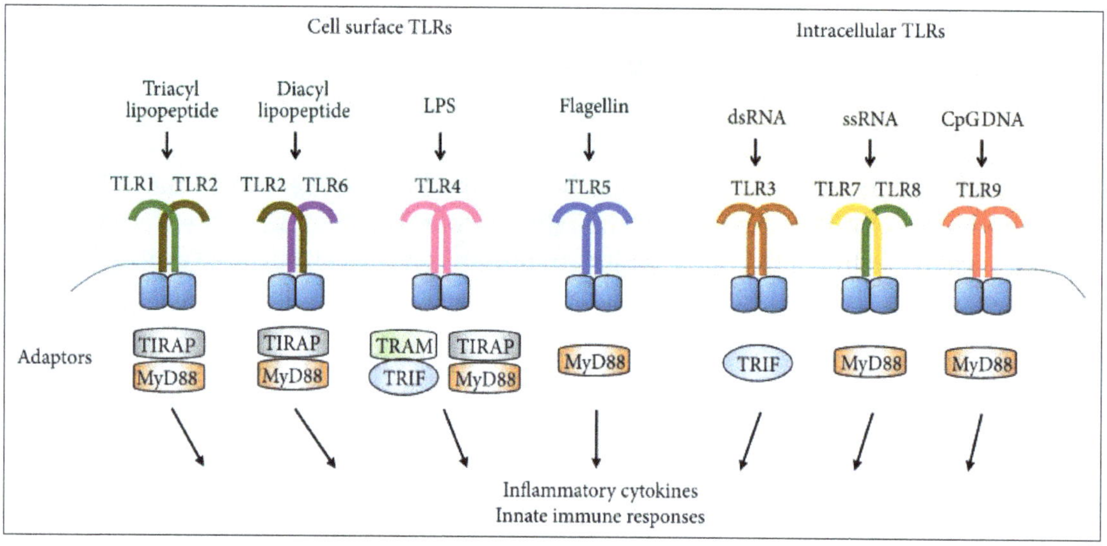

Function of the TLRs.

Table: Toll like receptors ligands and cells.

Receptor	Ligand	Microbes on which the ligand is present	Cells
TLR1	Lipopeptides	Gram-negative bacteria, my-cobacteria	MC, MA, DC
TLR2	Lipoteichoic acid eptidoglycan Zymosan Lipo arabino-mannan	Both Gram positive and Gram negative bacteria, some fungi, mycobacteria, Trypanosoma cruzi	MC, HA, PMN, DC, Mast Cells, FDC
TLR3	dsDNA	Virus	MA, DC
TLR4	LPS	Gram negative bacteria, some fungi	MA, PMC, DC, mast cells, FDC
TLR5	Flagellin	Bacteria	MC, MA, DC
TLR6	Lipopeptides	Bacteria	MC, MA, Mast cells

TLR7	ssDNA	Viruses	MC, MA, DC
TLR8	ssDNA	Viruses	MC, MA, DC
TLR9	dsDNA	Bacteria	MC, MA, DC

MC=Monocytes; MA= Macrophages; DC=Myeloid and plasmacytoid dendritic cells;

PMN=Polymorphonuclear cells; FDC=Follicular dendritic cells.

TLR2 can form heterodimers with TLR1 or TLR6 with tri-acyl proteins and di-acyl proteins as ligands TLR3, TLR7, TLR8, TLR9 are intracellular TLRs found on endosomes.

Some Additional Information on TLRs

TLR10 (also known as CD290) has been reported; its ligands are unknown but its cellular distribution is plasmacytoid dendritic cells. TLRs have extracellular domain rich in leucine rich repeats and a cytoplasmic domain that is involved in the signal transduction process. These germline coded molecules detect invading pathogens and trigger expression of cytokines and costimulating molecules. The best characterized member is TLR4. Apart from LPs, it also binds to host proteins during damage, infection and inflammation like HSP-60 and a fibronectin variant. The recognition of LPs by TLR4 also requires CD14 as a coreceptor and LPs binding protein from serum which brings LPs to CD14.

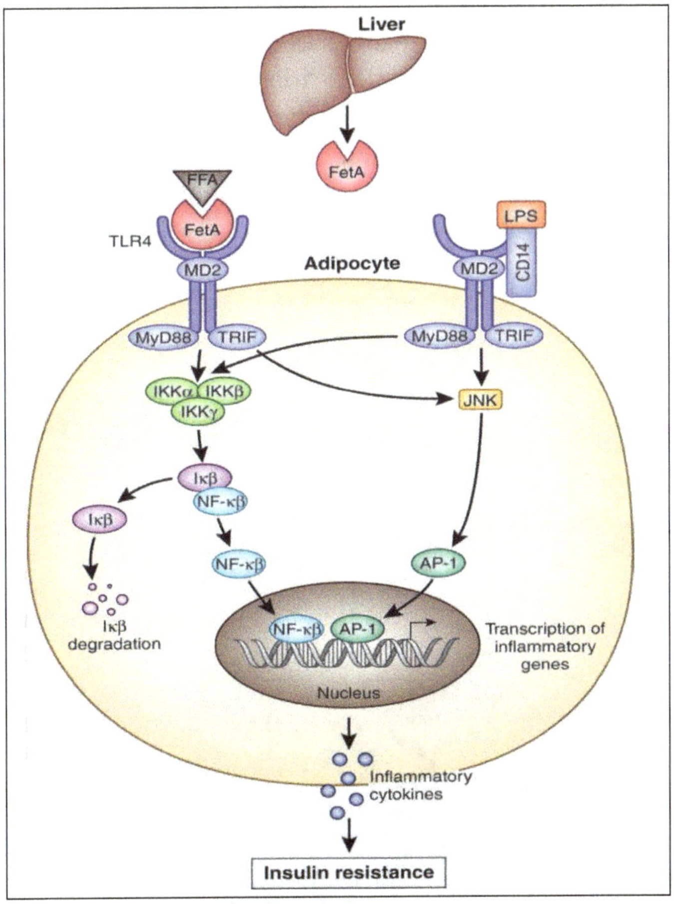

Recognition of LPS by TLR4.

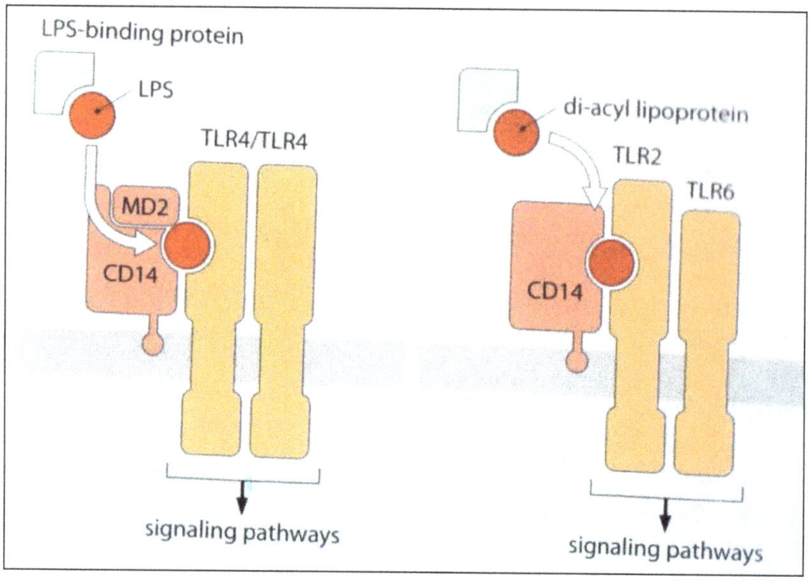

Activation of macrophages by LPS.

Mannose Receptor (MRs) (CD206)

It is a Ca^{2+} dependent 180 kD integral protein, has 8 carbohydrate lectin domains which recognizes mannosyl/fucosyl recognition patterns on pathogens entering through mucosal surfaces. It also has a terminal lectin domain which recognizes sulfated carbohydrate groups Macrophages (throughout the body), endothelial cells and dendritic cells have these receptors MRs link the innate immune response with acquired immunity as the pathogens are internalized and degraded in endosomes and displayed on antigen presenting cells. The receptor also binds host proteins such as myeloperoxidase, lysosomal hydrolases and some hormones.

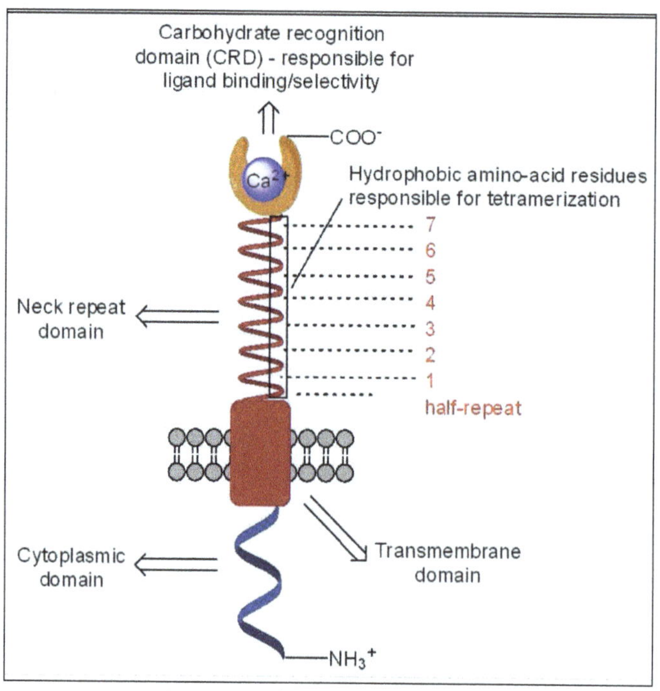

Structure of Mannose Receptors.

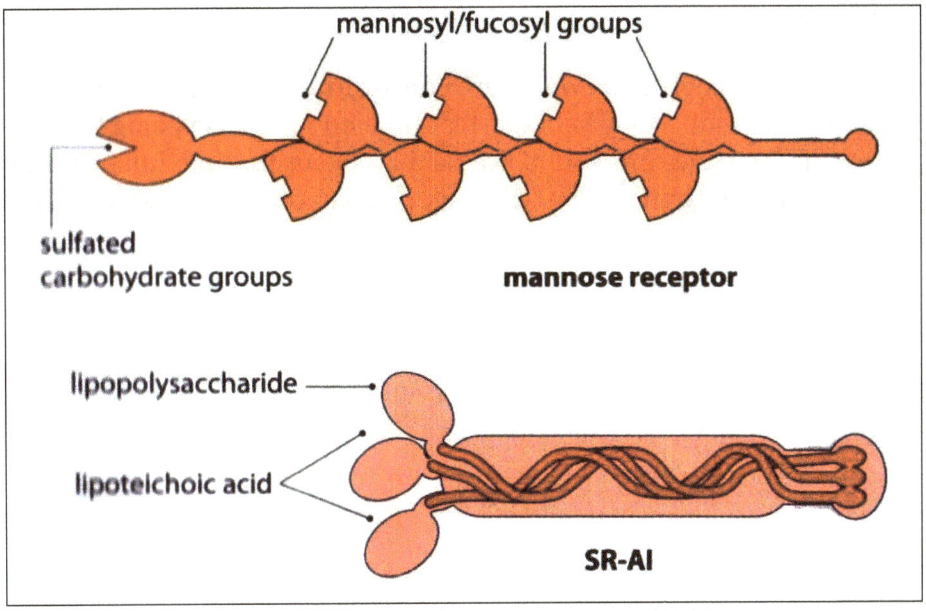

Mannose receptor and scavenger receptor SR AI.

Some other Important Receptors

Receptor	Ligand	Location
NOD-like receptors Scavenger receptors	Peptidoglycans Muramyldipeptides	Macrophages, Dendritic cells, Epithelial cells
	Lipo polysaccharides, Lipoteichoic acid	Macrophages, Dendritic cells, Some endothelial cells
Siglecs (Family of 12xproteins including siglec-1 (CD169), Siglec-2 (CD-22), Siglec-3 (CD- 33)	Sialic Acid	Macrophages
Dectins DC-Sign (Dendritic cell specific ICAM 3-grabbing non-integrin)a	Carbohydrates	Macrophages and dendritic cells
	Mannose	Dendritic cells, some macrophages
CD14	Lipopolysaccharides	Macrophages

Soluble Pattern Recognition Receptors

Many soluble proteins recognize PAMPs (Pathogen associated molecular patterns).

Collectins	Belongs to collagen containing C-type lectin superfamily; Acute phase proteins and mannose binding lectin (MBL) exemplify this case.
Ficollins	Have a collagen domain as well as fibrinogen domain. These recognize N-acetyl glucosamine and activate complement pathway.
Galectins	Bind to glycans and directly kill bacteria.

Many of this carbohydrate recognizing protein are part of the innate immunity.

Complement System

It consists of 20 soluble glycoproteins which are constitutively present in blood and other body fluids. Many of these are produced by hepatocytes and monocytes. On being triggered, a cascade of reactions occur which enables protection against pathogens through inflammation, enhanced phagocytosis or lysis of pathogenic cells. The triggering can occur through three different pathways.

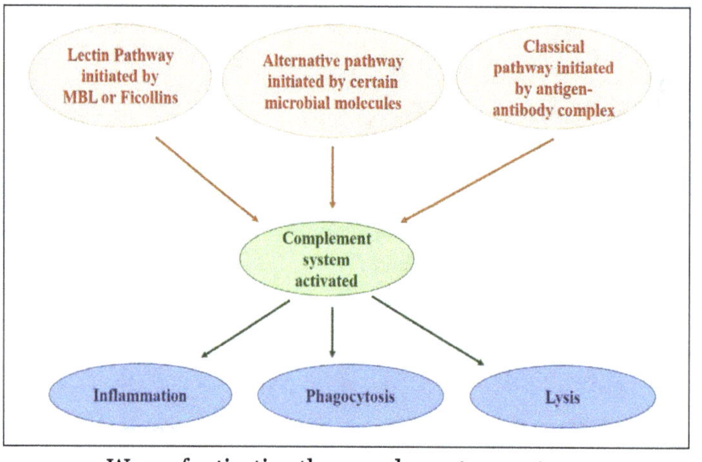

Ways of activating the complementary system.

Acute Phase Proteins

These include C-reactive Proteins (CRPs), Complement components, MBL, metal binding proteins and protease inhibitors. These are mostly produced by liver either de novo (e.g. CRP) or their low level increases (e.g. fibrinogen) rapidly. These are particularly important in defense against bacteria and protozoa. Activated macrophages and NK cells release cytokines (IL-6 being very important) and this leads to hepatocytes responding by producing acute phase proteins.

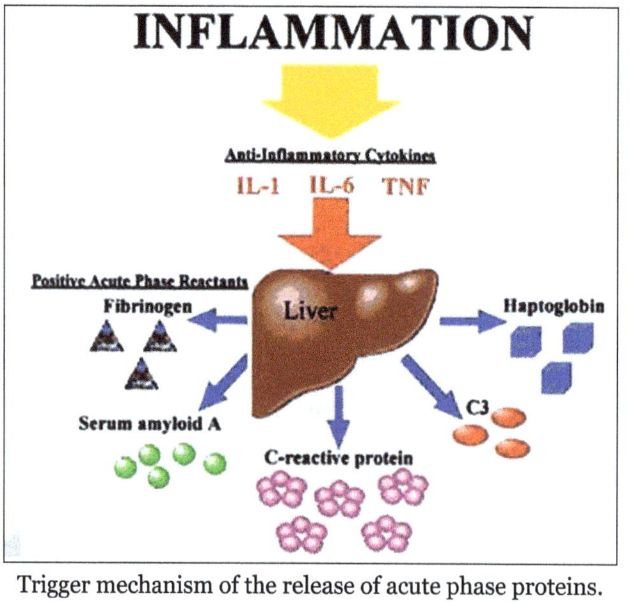

Trigger mechanism of the release of acute phase proteins.

Overall, acute phase proteins limit tissue damage by:

- Microbial infections,
- Trauma,
- Malignancy,
- Diseases like rheumatoid arthritis.

These proteins are important in tissue repair as well.

Various Acute Phase Proteins

Acute Phase Protein	Function
C-reactive protein (CRP; so named as it reacts with C-protein of pneumococcus)	Opsonin, Activates complement system.
Serum amyloid A (SAA)	Opsonin, Activates complement system.
Serum amyloid P (SAP)	Recognizes products of tissue breakdown including amyloid fibres.
MBL	Binds to mannose on bacterial surface, binds to phagocytes, Activates complement system.
Complement components	Chemotaxis, opsonization and lysis.
Metal binding proteins	Removal of metal ions essential for bacterial growth.
Fibrinogen	Clot formation.
Protease inhibitors (α-antitrypsin and α- antichymotrypsin)	Inhibits corresponding proteases.

Many acute phase proteins occur as similar pentagonal subunit association and hence are classified as Pentraxins. Examples are: CRP, SAA, SAP, All Pentraxins are Ca^{2+} dependant carbohydrate binding proteins.

Cytokines

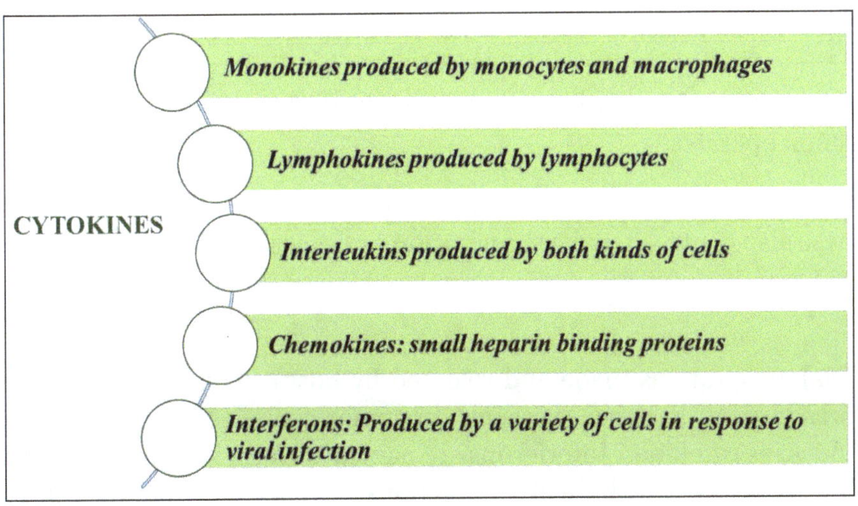

It is better to take the above not as a classification but a definition of terms as there is some overlap among those:

- Interleukins: 35 interleukins have been described, IL1-IL12 are better characterized,
- Some are also produced by lymphocytes and are thus lymphokines as well,
- These can be placed in four families based upon structural similarities.

IL1 family	IL1, IL18, IL33
IL2 family	IL2, IL2, IL15
IL10 family	IL10, IL19, IL20, IL22, IL24, IL26
IL12 family	IL12, IL23, IL27, IL35

Like other cytokines, these have diverse functions and highlight the overlap between innate and acquired immunity.

Other Important Cytokines

Chemokines

- More than 50 identified so far, these are 8-10 kDa peptides responsible for chemotaxis of lymphocytes, monocytes and neutrophils.
- There are conserved cysteine residues, their position and amino acid sequences form the basis of classes of chemokines.

Class	Examples	Attracts for activation
CXC (α)	IL-8, NAP-2, MIP-16	T-cells, PMNs, CD-8 T cells
CC (β)	MCP-1, Rantes	Memory T cells, monocytes and memory T_n cells
C (γ)	Lymphotactin	Lymphocytes
CX_3C (δ)	Fractalkine	Lymphocytes, monocytes, NK cells

- Receptors for chemokines are integral membrane proteins and G-protein mediated signal transduction operates.
- Hence these molecules activate and direct effector cells to the infection or tissue damage sites. These also control leukocyte migration in tissues.

Interferons

Interferons (IFNs) are proteins made and released by host cells in response to the presence of pathogens such as viruses, bacteria, parasites, or tumor cells. IFNs belong to the large class of glycoproteins known as cytokines. Interferons are named after their ability to "interfere" with viral replication within host cells. IFNs have other functions: they activate immune cells, such as natural killer cells and macrophages, they increase recognition of infection or tumor cells by up-regulating antigen presentation to T lymphocytes, and they increase the ability of uninfected host cells to resist new infection by virus. Certain symptoms, such as aching muscles and fever, are related to the production of IFNs during infection.

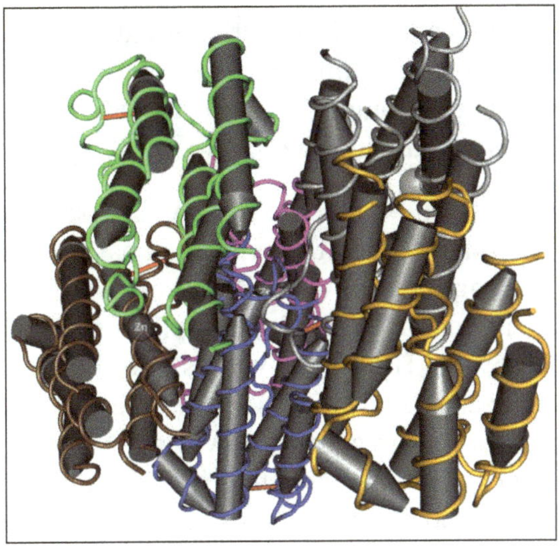

Interferon: The molecular structure of human interferon-alpha.

About ten distinct IFNs have been identified in mammals; seven of these have been described for humans. They are typically divided among three IFN classes: type I IFN, type II IFN, and type III IFN. IFNs belonging to all IFN classes are very important for fighting viral infections.

Based on the type of receptor through which they signal, human interferons have been classified into three major types:

- Interferon type I: All type I IFNs bind to a specific cell surface receptor complex, known as the IFN-α receptor (IFNAR) that consists of IFNAR1 and IFNAR2 chains. The type I interferons present in humans are IFN-α, IFN-β and IFN-ω.
- Interferon type II: These bind to IFNGR that consist of IFNGR1 and IFNGR2 chains. In humans this is IFN-γ.
- Interferon type III: These signal through a receptor complex consisting of IL10R2 (also called CRF2-4) and IFNLR1 (also called CRF2-12). Acceptance of this classification is less universal than that of type I and type II, and unlike the other two, it is not currently included in *Medical Subject Headings*.

Effects of Interferons

All interferons share several common effects; they are antiviral agents and can fight tumors. As an infected cell dies from a cytolytic virus, viral particles are released that can infect nearby cells. In addition, interferons induce production of hundreds of other proteins—known collectively as interferon-stimulated genes (ISGs)—that have roles in combating viruses. They also limit viral spread by increasing p53 activity, which kills virus-infected cells by promoting apoptosis. The effect of IFN on p53 is also linked to its protective role against certain cancers. Another function of interferons is to upregulate major histocompatibility complex molecules, MHC I and MHC II, and increase immunoproteasome activity. Interferons, such as interferon gamma, directly activate other immune cells, such as macrophages and natural killer cells. Interferons can inflame the tongue and cause dysfunction in taste bud cells, restructuring or killing taste buds entirely.

By interacting with their specific receptors, IFNs activate signal transducer and activator of transcription (STAT) complexes. STATs are a family of transcription factors that regulate the expression of certain immune system genes. Some STATs are activated by both type I and type II IFNs. However, each IFN type can also activate unique STATs.

STAT activation initiates the most well-defined cell signaling pathway for all IFNs, the classical Janus kinase-STAT (JAK-STAT) signaling pathway. In this pathway, JAKs associate with IFN receptors and, following receptor engagement with IFN, phosphorylate both STAT1 and STAT2. As a result, an IFN-stimulated gene factor 3 (ISGF3) complex forms—this contains STAT1, STAT2 and a third transcription factor called IRF9—and moves into the cell nucleus. Inside the nucleus, the ISGF3 complex binds to specific nucleotide sequences called IFN-stimulated response elements (ISREs) in the promoters of certain genes, known as IFN stimulated genes ISGs. Binding of ISGF3 and other transcriptional complexes activated by IFN signaling to these specific regulatory elements induces transcription of those genes. Interferome is a curated online database of ISGs. Additionally, STAT homodimers or heterodimers form from different combinations of STAT-1, -3, -4, -5, or -6 during IFN signaling; these dimers initiate gene transcription by binding to IFN-activated site (GAS) elements in gene promoters. Type I IFNs can induce expression of genes with either ISRE or GAS elements, but gene induction by type II IFN can occur only in the presence of a GAS element.

In addition to the JAK-STAT pathway, IFNs can activate several other signaling cascades. Both type I and type II IFNs activate a member of the CRK family of adaptor proteins called CRKL, a nuclear adaptor for STAT5 that also regulates signaling through the C3G/Rap1 pathway. Type I IFNs further activate p38 mitogen-activated protein kinase (MAP kinase) to induce gene transcription. Antiviral and antiproliferative effects specific to type I IFNs result from p38 MAP kinase signaling. The phosphatidylinositol 3-kinase (PI3K) signaling pathway is also regulated by both type I and type II IFNs. PI3K activates P70-S6 Kinase 1, an enzyme that increases protein synthesis and cell proliferation; phosphorylates of ribosomal protein s6, which is involved in protein synthesis; and phosphorylates a translational repressor protein called eukaryotic translation-initiation factor 4E-binding protein 1 (EIF4EBP1) in order to deactivate it.

Some other Important Cytokines

- Migration Inhibition Factor (MIF): Released by activated T-cells, it inhibits the migration of macrophages and thus enhances their action at the site of inflammation.
- Tumour necrosis factor (TNF) and lymphotoxin (LT): Both are structurally related and in fact LT is also called TNFβ. TNFα is produced by macrophages and some other cells whereas TNFβ is not a part of the innate immunity as it is produced by Tc cells. Their actions include:
    - Enhancing adhesion of leukocytes to vascular endothelium.

- Mobilization of fat resulting in cachexia (washing).
- Apoptosis by activating caspases.

- Transforming growth factor-β (TGF-β): These are group of 5 cytokines released by many cells including macrophages and platelets. Actions include increasing production of the extracellular matrix protein and controlling proliferation of B- and T-cells.
- CSFs including granulocyte-monocyte CSF (GM-CSF), granulocyte CSF (G-CSF) and monocyte CSF (M-CSF).

These are involved in the development, differentiation and expansion of cells of myeloid series.

Cells of the Innate Immune System

Phagocyte

Phagocyte is a type of cell that has the ability to ingest, and sometimes digest, foreign particles, such as bacteria, carbon, dust, or dye. It engulfs foreign bodies by extending its cytoplasm into pseudopods (cytoplasmic extensions like feet), surrounding the foreign particle and forming a vacuole. Poisons contained in the ingested bacteria cannot harm the phagocyte so long as the bacteria remain in the vacuole; phagocyte enzymes are secreted into the vacuole in which digestion takes place. In the blood, two types of white blood cells, neutrophilic leukocytes (microphages) and monocytes (macrophages), are phagocytic. Neutrophils are small, granular leukocytes that quickly appear at the site of a wound and ingest bacteria. Monocytes are larger, with a large, kidney-shaped nucleus; they appear about three days after infection and scavenge for bacteria, foreign particles, dead cellular material, and protozoa. Most phagocytic activity takes place outside the vascular system, among the cells. For example, foreign material in the lymph system is phagocytosed by fixed cells in the lymph nodes; similarly, the vascular system is cleansed by fixed cells in the spleen, liver, and bone marrow that engulf aged red blood cells and foreign bodies. Special cells in the lungs ingest dust particles. Fixed macrophages in loose connective tissue, as in spleen or bone marrow, are capable of rounding up and becoming free macrophages under the proper stimulus (e.g., inflammation).

Natural Killer Cells

Natural Killer (NK) Cells are lymphocytes in the same family as T and B cells, coming from a common progenitor. However, as cells of the innate immune system, NK cells are classified as group I Innate Lymphocytes (ILCs) and respond quickly to a wide variety of pathological challenges. NK cells are best known for killing virally infected cells, and detecting and controlling early signs of cancer. As well as protecting against disease, specialized NK cells are also found in the placenta and may play an important role in pregnancy.

NK cells were first noticed for their ability to kill tumour cells without any priming or prior activation (in contrast to cytotoxic T cells, which need priming by antigen presenting cells). They are named for this 'natural' killing. Additionally, NK cells secrete cytokines such as IFNγ and TNFα, which act on other immune cells like Macrophage and Dendritic cells to enhance the immune response.

While on patrol NK cells constantly contact other cells. Whether or not the NK cell kills these cells depends on a balance of signals from activating receptors and inhibitory receptors on the NK cell surface. Activating receptors recognise molecules that are expressed on the surface of cancer cells and infected cells, and 'switch on' the NK cell. Inhibitory receptors act as a check on NK cell killing. Most normal healthy cells express MHC I receptors which mark these cells as 'self'. Inhibitory receptors on the surface of the NK cell recognise cognate MHC I, and this 'switches off' the NK cell, preventing it from killing. Cancer cells and infected cells often lose their MHC I, leaving them vulnerable to NK cell killing. Once the decision is made to kill, the NK cell releases cytotoxic granules containing perforin and granzymes, which leads to lysis of the target cell.

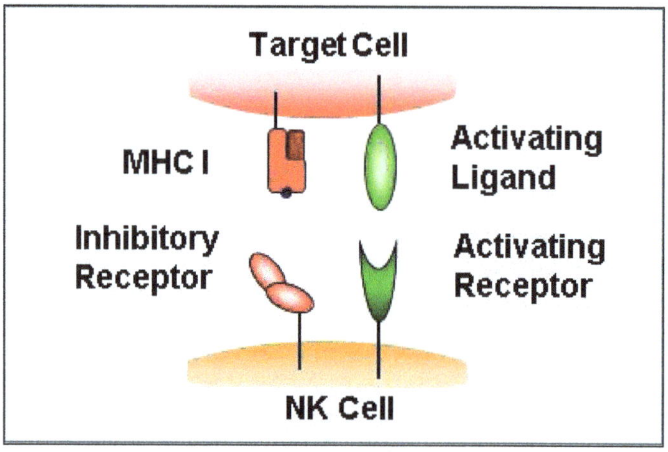

The genes for both MHC I and the NK cell inhibitory receptors which recognise them vary a lot between individuals. The versions (or alleles) of these genes a person carries have been linked to their ability to fight HIV infection and their risk of some autoimmune diseases. NK cell varieties also change with age and are affected by chronic viral infections such as cytomegalovirus (CMV).

Because of their ability to kill tumour cells, NK cells are an attractive target for cancer immunotherapies. Some therapeutic monoclonal antibodies rely on NK cell killing. Researchers are also developing treatments to activate NK cells using small molecules or cytokines, and even testing genetically modified living NK cells as therapies.

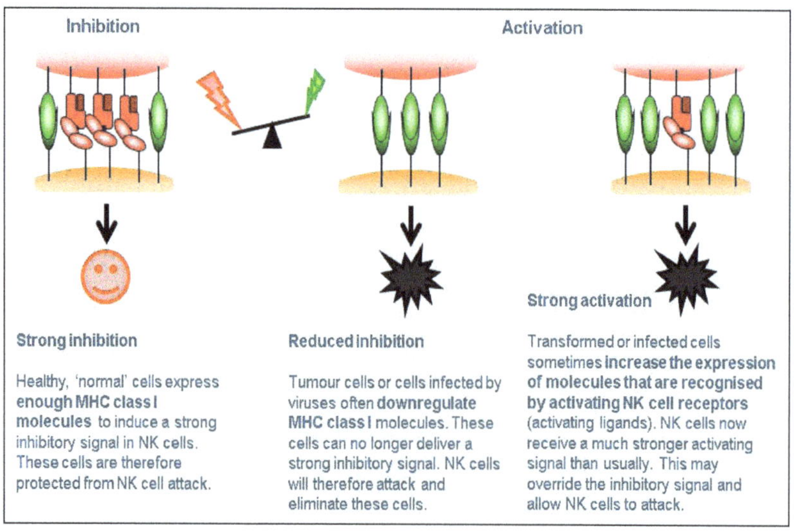

Mast Cells vs. Basophils

Both mast cells and basophils share a number of characteristics in relation to their function and development. Like mast cells, basophils are also involved in various allergic and inflammatory diseases. This is because the two produce high amounts of mediators like histamine that play a central role in inflammatory responses.

The two have been shown to share a number of features such as metachromatic granules and high-affinity receptors that enable them to identify and mount a response against invading organisms (bacteria and multicellular parasites like helminths).

Like mast cells, basophils are also referred to as effector cells. As such, they are capable of producing various bioactive molecules (e.g. cytokines, proteases, and chemokines etc.). These molecules are actively involved in several processes that are directed towards protecting the host. For instance, histamine and cytokines play an important role in inflammation responses as well as in reactions against infections brought about by parasites,

While mast cells and basophils share a number of features, differences between the two are evident in their respective life span, where they are located in the body as well as proliferation. Mast cells arise from the bone marrow and are released into the bloodstream as immature cells only to mature in peripheral tissue. This is different from basophils that mature in the bone marrow and circulate in peripheral blood as mature cells.

Unlike mast cells, which are long-lived, the life span of basophils is significantly shorter given that they can only survive for a few days (about 3 days compared to mast cells that can survive for a few months). This characteristic of basophils may be attributed to the fact that they do not proliferate once they mature. While mast cells only differentiate in tissue, they are capable of proliferating in these sites and therefore continue renewing themselves.

Dendritic Cell

Dendritic cells are a type of antigen-presenting cell (APC) that forms an important role in the adaptive immune system. The main function of dendritic cells is to present antigens and the cells are therefore sometimes referred to as "professional" APCs.

In addition, only the dendritic cells have the capacity to induce a primary immune response in the inactive or resting naïve T lymphocytes. To do this, the dendritic cells capture the antigens from invading bodies, which they process and then present on their cell surface and presented, along with the necessary accessory or co-stimulation molecules.

Dendritic cells also contribute to the function of B cells and help maintain their immune memory. Dendritic producing cytokines and other factors that promote B cell activation and differentiation. After an initial antibody response has occurred due to an invading body, dendritic cells found in the germinal centre of lymph nodes seem to contribute to B cell memory by forming numerous antibody-antigen complexes. This is to provide a long lasting source of antigen that the B cells can take up themselves and present to T cells.

Dendritic cells are found in tissue that has contact with the outside environment such as the over the skin (present as Langerhans cells) and in the linings of the nose, lungs, stomach and intestines.

Immature forms are also found in the blood. Once activated, dendritic cells move to the lymph tissue to interact with to interact with T cells and B cells and help shape the adaptive immune response. During development, they develop branched projections called "dendrites", which is why the cells are so named.

Other Cells which have Roles in Innate Immunity

- Erythrocytes: Bind and remove small immune complexes,
- Eosinophiles: Granular leukocytes which release toxic proteins which kill parasites,
- Platelets: release mediators that activate complement cells,
- Epithelial cells: The role of these in releasing AMPs has already been mentioned.

Humoral and Cellular Immunity

Humoral Immunity means the protection offered by antibodies. Antibodies themselves are produced by B-lymphocytes. Cellular Immunity means the immune mechanism is mediated by T-cells which are another class of lymphocytes.

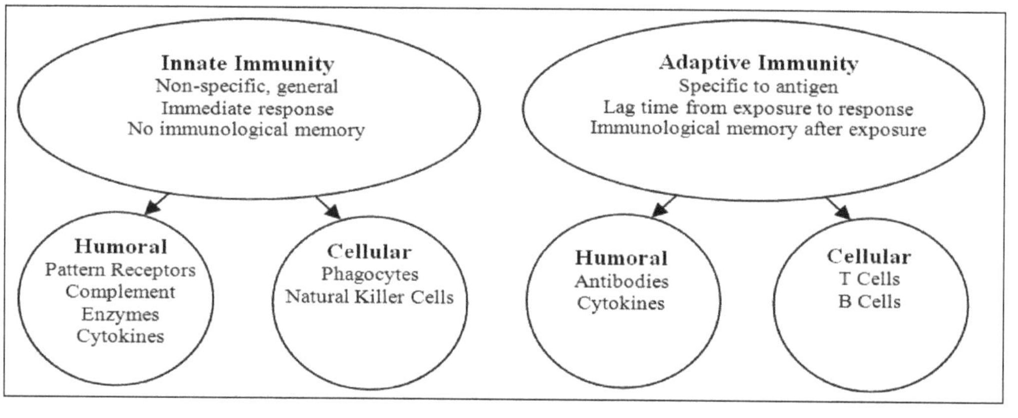

It is the right time to talk of the relative role of the innate and the acquired or adaptive immunity. With regards to the latter, we need to look at the broad level of how these co-operate.

Humoral and Cell-Mediated Immune Responses

The immune system distinguishes two groups of foreign substances. One group consists of antigens that are freely circulating in the body. These include molecules, viruses, and foreign cells. A second group consists of self-cells that display aberrant MHC proteins. Aberrant MHC proteins can originate from antigens that have been engulfed and broken down (exogenous antigens) or from virus-infected and tumor cells that are actively synthesizing foreign proteins (endogenous antigens). Depending on the kind of foreign invasion, two different immune responses occur:

- The humoral response (or antibody-mediated response) involves B cells that recognize antigens or pathogens that are circulating in the lymph or blood ("humor" is a medieval term for body fluid). The response follows this chain of events:

- Antigens bind to B cells.
- Interleukins or helper T cells costimulate B cells. In most cases, both an antigen and a costimulator are required to activate a B cell and initiate B cell proliferation.
- B cells proliferate and produce plasma cells. The plasma cells bear antibodies with the identical antigen specificity as the antigen receptors of the activated B cells. The antibodies are released and circulate through the body, binding to antigens.
- B cells produce memory cells. Memory cells provide future immunity.

- The cell-mediated response involves mostly T cells and responds to any cell that displays aberrant MHC markers, including cells invaded by pathogens, tumor cells, or transplanted cells. The following chain of events describes this immune response:
    - Self-cells or APCs displaying foreign antigens bind to T cells.
    - Interleukins (secreted by APCs or helper T cells) costimulate activation of T cells.
    - If MHC-I and endogenous antigens are displayed on the plasma membrane, T cells proliferate, producing cytotoxic T cells. Cytotoxic T cells destroy cells displaying the antigens.
    - If MHC-II and exogenous antigens are displayed on the plasma membrane, T cells proliferate, producing helper T cells. Helper T cells release interleukins (and other cytokines), which stimulate B cells to produce antibodies that bind to the antigens and stimulate nonspecific agents (NK and macrophages) to destroy the antigens.

Structure of Antibodies and Immunoglobulin Domains

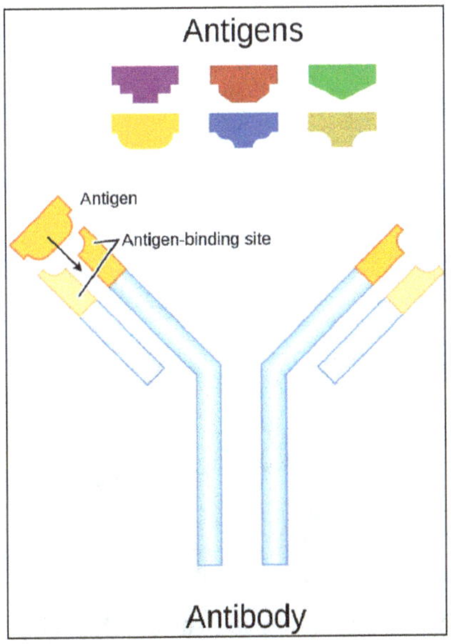

Each antibody binds to a specific antigen; an interaction similar to a lock and key.

Each antibody molecule consists of four polypeptides, two heavy chains, and two light chains joined to form a "Y" shaped molecule. The chains are connected by disulfide bonds. There are five types of Ig heavy chains (in mammals) denoted by the Greek letters: α, δ, ε, γ, and μ. There are two types of Ig light chains (in mammals), which are called lambda (λ) and kappa (κ).

Antibodies consist of a variable region and a constant region. The variable region changes to various structures depending on differences in antigens. The variability is expressed in the amino acid sequence in the tips of the "Y". The constant region determines the mechanism used to destroy the antigen.

Antibody Polymers and Antibody Binding to Antigens

The antigen-antibody reaction is widely used in laboratory diagnostics, including immunohematology. It is a reversible chemical reaction:

Antigen + antibody ⇄ antigen-antibody complex

The forces connecting the antigen-antibody are not sturdy valence bonds, but weaker bonds that are fittingly named "weak interactions."

The part of the antibody that interacts with the antigen is in the Fab portion of the antibody molecule and is assembled from the hypervariable regions of the heavy and light chains located at the tips of the "Y." The binding between this site and the antigen takes place with the following characteristics and processes:

- The bonds holding the antigen to the antibody are non-covalent/reversible in nature. They may be hydrogen bonds, electrostatic bonds, or Van der Waals forces.
- Usually, there are multiple bond formations observed to ensure relatively tight binding between antibody and antigen.
- The specific binding between the antigenic determinant on the cell (known as epitope) and the antigen combining site (paratope) on the antibody involves very small portions of the molecules, usually comprising only a few amino acids.

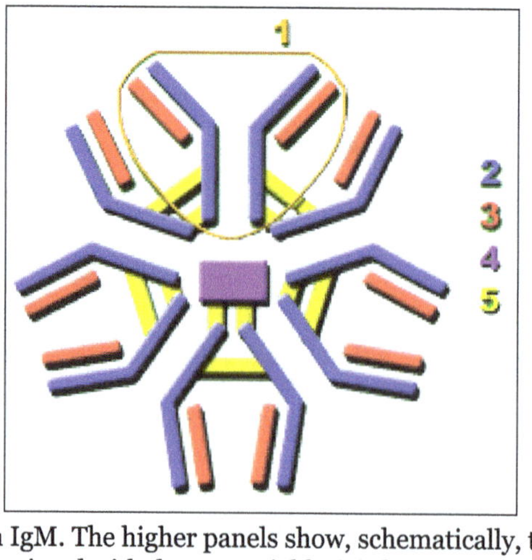

"Secreted mammalian IgM. The higher panels show, schematically, the folding pattern of the constant (C) associated with degree variable (V) domains of a lightweight chain."

- These sites are critical in antigen-antibody reactions as specific binding has to overcome the natural repulsion between the two molecules.
- When the two sites come together they are first attracted to each other by ionic and hydrophobic forces.
- These forces help them overcome their hydration energies and allow for the expulsion of water molecules as the two parts approach each other.
- This attraction becomes even stronger when Van der Waals forces are employed later on to bring the epitope and paratope even closer.

The Function of Antibodies

- IgG:
    - Provides long term protection because it persists for months or years after initial exposure to the antigen.
    - Protects against bacteria and viruses.
    - Neutralizes bacterial toxins.
    - Triggers compliment protein systems and bind antigens to enhance the effectiveness of phagocytosis.
- IgA:
    - Binds antigens on microbes before they invade tissues.
    - Aggregates antigens and keeps them within mucosal secretions, so that the antigen will be expelled with the secretion (e.g. coughing, sneezing).
    - Defends mucosal surfaces such as the intestines, nose, and lungs.
- IgM:
    - Involved in the antigens on the surface of red blood cells that determine the ABO blood group.
    - Enhances the ingestion of cells via phagocytosis.
- IgE:
    - Binds to mast cells and basophils which participate in the immune response.
    - Protects against parasites.
- IgD:
    - Present on the surface of B cells.
    - Plays a role in the induction of antibody production.

Classes of Antibodies (Immunoglobulin)

The Five immunoglobulin (Ig) classes					
	Igm pentamer	IgG monomer	Secretory iga dimer	IgE monomer	IgD monomer
	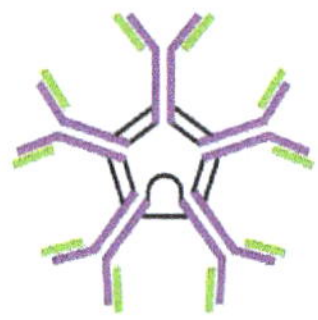	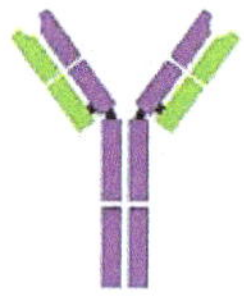	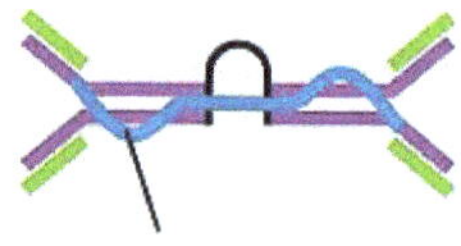 Secretory Components	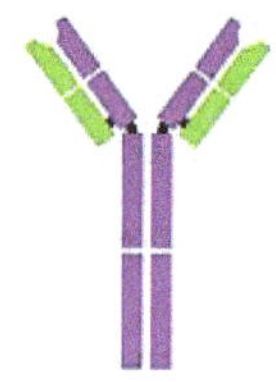	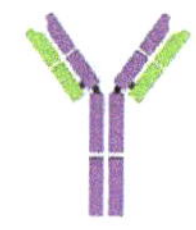
Heavy chains	μ	γ	α	ε	δ
Number of antigen binding sites	10	2	4	2	2
Moleculer weight (daltons)	900,000	150,000	385,000	200,000	180,000
Percentage of total antibody in serum	6%	80%	13%	0.002%	1%
Crosses placenta	No	Yes	No	No	No
Fixes complement	Yes	yes	No	No	No
Fc bind to		Phagocytes		Mast cells and basophils	
Function	Main antibody of primary responses, best at fixing complement; the monomer from of IgM serves as the B cell receptor.	Main blood antibody of secondary responses, neutralizes toxin opsonization.	Secreted into mucus, tears, saliva, colostrum.	Antibody of allergy and antiparasitic activity.	B cell receptor.

Autoimmune Disease

Autoimmune diseases can affect the human immune system, so the primary goal is to create self-immunity. However, if a person's body hasn't developed an adequate immune system, some autoimmune diseases can appear, such as rheumatic fever, rheumatoid arthritis, ulcerative colitis, myasthenia gravis, Guillain-Barré syndrome, Reiter's syndrome, reactive arthritis, or type 1 diabetes mellitus.

Similarities between Humoral Immunity and Cell Mediated Immunity

- Humoral immunity and cell mediated immunity are two types of adaptive immunity.
- Both humoral immunity and cell mediated immunity produce a specific immune response to a particular pathogen.

Difference between Humoral and Cell Mediated Immunity

- Definition:
    - Humoral Immunity: Humoral immunity refers to a component of the adaptive immunity where B cells secrete antibodies, which circulate in the blood as a soluble protein.
    - Cell Mediated Immunity: Cell mediated immunity refers to the other component of the adaptive immunity, which is mediated by the activated, antigen-specific T cells.
- Main Cells:
    - Humoral Immunity: The humoral immunity is mediated by B cells.
    - Cell Mediated Immunity: The cell mediated immunity is mediated by T cells.
- Cell Types:
    - Humoral Immunity: Humoral immunity is mediated by T cells, B cells, and macrophages.
    - Cell Mediated Immunity: Cell mediated immunity is mediated by helper T cells, cytotoxic T cells, natural killer cells, and macrophages.
- Action:
    - Humoral Immunity: The humoral immunity acts on the extracellular microbes and their toxins.
    - Cell Mediated Immunity: The cell mediated immunity acts on intracellular microbes such as viruses, bacteria, and parasites and tumor cells.
- Receptors:
    - Humoral Immunity: The BCR receptors are involved in the humoral immunity.
    - Cell Mediated Immunity: The TCR receptors are involved in the cell mediated immunity.
- Accessory Surface Molecules:
    - Humoral Immunity: The Igα, Igβ, CD40, CD21, and Fc receptors are the accessory receptors of the humoral immunity.
    - Cell Mediated Immunity: The CD2, CD3, CD4, CD8, CD28, and integrins are the accessory receptors of the cell mediated immunity.
- Role of MHC Molecules:
    - Humoral Immunity: The unprocessed antigens are recognized by the humoral immunity.
    - Cell Mediated Immunity: The antigens are processed and presented by MHC complexes in the cell mediated immunity.

- Secretion:
    - Humoral Immunity: The plasma B cells secrete antibodies in the humoral immunity.
    - Cell Mediated Immunity: The T cells secrete cytokines.
- Onset:
    - Humoral Immunity: The humoral immune response is rapid.
    - Cell Mediated Immunity: The cell-mediated immune response is a delayed type of hypersensitivity.
- Tumor Cells and Transplants:
    - Humoral Immunity: The humoral immunity does not act on the tumor cells and transplants.
    - Cell Mediated Immunity: The cell mediated immunity acts on tumor cells and transplants.

Immunization

Immunization is a process by which resistance to disease is acquired or induced in plants and animals. Immunization may occur naturally, as when a person is exposed unintentionally to a pathogen (any infectious agent), or it may be brought about artificially through a vaccine. In either case, immunization provides resistance, or immunity, to a particular pathogen by means of antibody proteins that are targeted to eliminate that pathogen from the body. These antibodies do not react to the entire pathogen but only to a specific part of it, which is called an antigen. An individual can acquire immunity for a specific pathogen passively or actively. In passive immunization a person receives antibodies or lymphocytes that have been produced by another individual's immune system; in active immunization the individual's own immune system is stimulated to produce antibodies and lymphocytes.

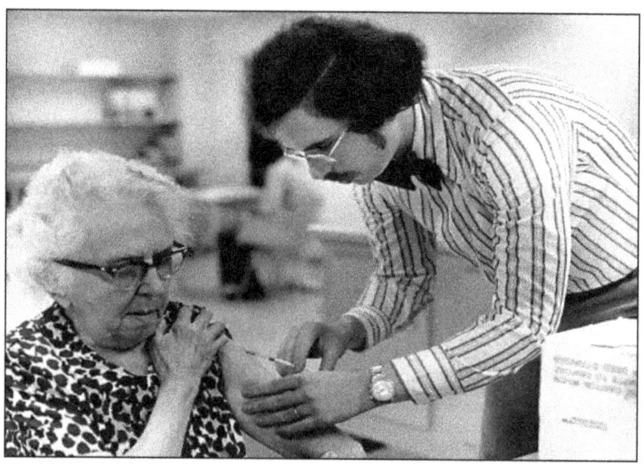

Immunization: Woman being immunized during the 1976 nationwide swine flu vaccination campaign. Fifty million Americans were vaccinated in a 10 week period.

Passive immunization imparts immediate, but not long-lasting, protection against a pathogen and may arise naturally, such as when a fetus receives antibodies from the mother across the placenta or when a breast-feeding infant ingests antibodies in the mother's milk. Passive immunization against a particular pathogen, such as the hepatitis B virus (HBV), also can be conferred artificially. A person lacking immunity to HBV can receive a preparation called immune serum globulin that contains antibodies formed against the virus. These antibodies are obtained from serum taken from animal or human donors who previously were infected by or immunized against the virus.

Active immunization stimulates the immune system to produce antibodies against a particular infectious agent. Active immunity can arise naturally, as when someone is exposed to a pathogen. For example, an individual who recovers from a first case of the measles is immune to further infection by the measles-causing virus, because the virus stimulates the immune system to produce antibodies that specifically recognize and neutralize the pathogen the next time it is encountered. Active immunization also can be artificially induced through vaccination. Vaccines are preparations containing antigens that stimulate an immune response without causing illness. The purpose of vaccination is to ensure that a large enough number of antibodies and lymphocytes capable of reacting against a specific pathogen or toxin are available before exposure to it occurs. Active immunization is often long-lasting and may be reactivated quickly by a recurrence of the infection or by revaccination.

References

- Immunology, earth-and-planetary-sciences: sciencedirect.com, Retrieved 27, March 2020
- Developmental-immunology: immunology.utoronto.ca, Retrieved 17, July 2020
- What-are-Dendritic-Cells, health: news-medical.net, Retrieved 05, August 2020
- Humoral-and-cell-mediated-immune-responses, the-immune-system-and-other-body-defenses, anatomy-and-physiology, study-guides: cliffsnotes.com, Retrieved 10, April 2020
- Primary-immune-response, humoral-and-cell-mediated-immunity, magazine: lecturio.com, Retrieved 14, June 2020
- Difference-Between-Humoral-and-Cell-Mediated-Immunity-320180727: researchgate.net, Retrieved 10, July 2020

Chapter 2

Antigens and Antibodies

Antigens are the substances which evoke an immune response in the body. Antibodies are specialized proteins produced by a body's immune system in response to the introduction of an antigen. There are different types of antigens and antibodies. This chapter is dedicated to the close study and examination of the subject matter for the reader's benefit.

Antigens

Antigen is any foreign substance that elicits an immune response (e.g. the production of specific antibody molecules) when introduced into the tissues of a susceptible animal, and is capable of combining with the specific antibodies formed. Antigens are generally of high molecular weight and are commonly proteins or polysaccharides. Polypeptides, lipids, nuclear acids and many other materials can also function as antigens.

Immune responses may also be generated against smaller substances, called haptens, if these are chemically coupled to a larger carrier protein. Common carrier proteins are bovine serum albumin (BSA), keyhole limpet hemocyanin (KLH) or other synthetic matrices.

A variety of molecules such as drugs, simple sugars, amino acids, small peptides, phospholipids or triglycerides may function as haptens. Thus, given enough time, just about any foreign substance will be identified by the immune system and evoke specific antibody production. However, this specific immune response is highly variable and depends much in part on the size, structure and composition of antigens. Antigens that elicit strong immune responses are said to be strongly immunogenic.

For efficient interaction to occur between the antigen and the antibody, the epitope (a part of an antigen that the specific antibody recognizes and binds to) must be readily available for binding. If the target molecule is denatured, e.g. through fixation, reduction, pH changes or during preparation for gel electrophoresis, the epitope may be altered and this may affect its ability to interact with an antibody.

For example, some antibodies are ineffective in western blot but very good in immunohistochemistry because in the latter procedure, a complex antigenic site might be maintained in the tissue, whereas in the former procedure the process of sample preparation alters the protein conformation sufficiently to destroy the antigenic site and hence eliminate antibody binding.

Thus, the epitope may be present in the antigen's native, cellular environment, or only exposed when denatured. In their natural form they may be cytoplasmic (soluble), membrane associated or secreted. The number, location and size of the epitopes depends on how much of the antigen is presented during the antibody making process.

If a gene product of interest is present in extremely low concentrations, one may choose to use known nucleotide sequence information to derive a corresponding peptide for generating sequence-specific antibodies (e.g. peptide specific antibodies). In some instances, peptide antigens have advantages over whole protein antigens in that the antibodies generated may be targeted to unique sequence regions. This is especially useful when investigating proteins that belong to families of high sequence homology.

Characteristics of a good antigen include:

- Areas of structural stability and chemical complexity within the molecule.
- Significant stretches lacking extensive repeating units.
- A minimal molecular weight of 8,000–10,000 Da, although haptens with molecular weights as low as 200 Da have been used in the presence of a carrier protein.
- The ability to be processed by the immune system.
- Immunogenic regions those are accessible to the antibody-forming mechanism.
- Structural elements those are sufficiently different from the host.
- For peptide antigens, regions containing at least 30% of immunogenic amino acids: lysine, arginine, glutamic acid, aspartic acid, glutamine, aspargine.
- For peptide antigens, significant hydrophilic or changed residues.

Molecular Structure of Antigens

At the molecular level, an antigen is characterized by its ability to be "bound" at the antigen-binding site of an antibody. Antibodies tend to discriminate between the specific molecular structures presented on the surface of the antigen. Antigens are usually proteins, peptides, or polysaccharides. This includes parts (coats, capsules, cell walls, flagella, fimbrae, and toxins) of bacteria, viruses, and other microorganisms. Lipids and nucleic acids are antigenic only when combined with proteins and polysaccharides. For example, the combination of lipids and polysaccharides are lipopolysaccharides (LPS), which are the primary component of gram negative bacterial endotoxin. LPS forms the cell wall of gram negative bacteria and causes a powerful immune response when bound. Cells present their immunogenic-antigens to the immune system via a major histocompatibility (MHC) molecule. Depending on the antigen presented and the type of the histocompatibility molecule, several types of immune cells can become activated due to an antigen.

Antigens have several structural components of interaction that may be bound by different classes of antibodies. Each of these distinct structural components is considered to be an epitope, also called an antigenic determinent. Therefore, most antigens have the potential to be bound by several distinct antibodies, each of which is specific to a particular epitope. The antigen binding receptor on an antibody is called a paratope, and is specific to the epitope of the antigen. Using the "lock and key" metaphor, the antigen itself can be seen as a string of keys – any epitope being a "key" – each of which can match a different lock.

Types of Antigens

Important microbial antigens are proteins or protein conjugates:

- Clostridial Toxins,
- Bacterial flagella,
- Viral Capsids,
- Protozoan cell membranes.

Other examples are:

- Snake venoms,
- Serum proteins,
- Milk proteins,
- Hormones.

Antibodies themselves constitute important antigens in experimental immunology. All antibodies are immunoglobulins and are proteins/glycoproteins.

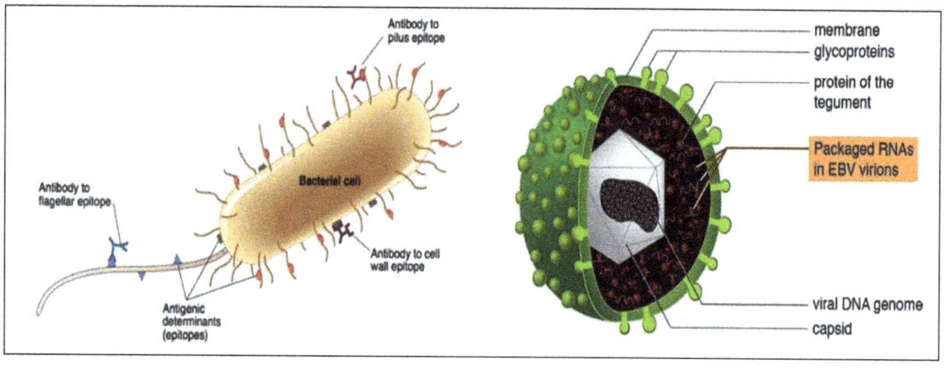

Cell wall lipopolysaccharides of gram negative bacteria are endotoxins.

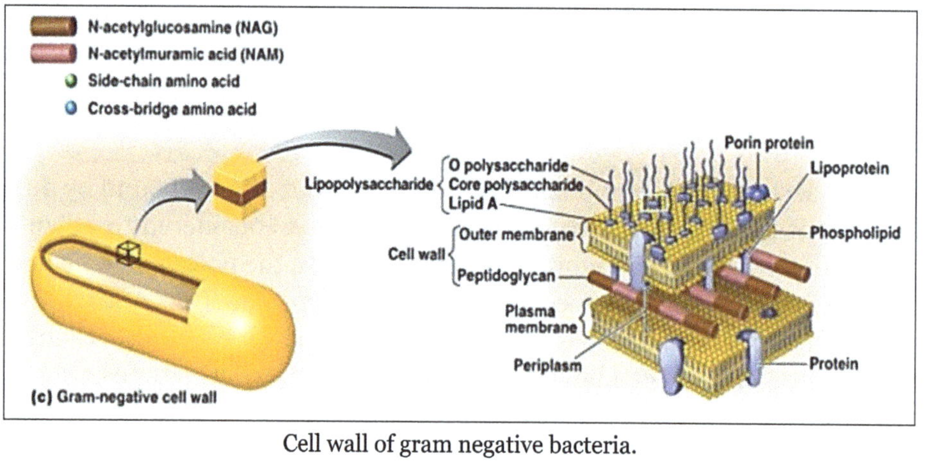

Cell wall of gram negative bacteria.

- Lipid A is an endotoxin.
- Lipopolysaccharide O is an antigen.

Some bacteria are enclosed by capsules which are either protein or polysaccharide in nature. Capsular antigens are called K-antigens. Pili or fimbriae are called F-antigens. Bacteria flagella consists of protein called flagellin and these antigens are called H-antigens. Proteins secreted by bacteria or obtained by autolysis of bacteria are called endotoxins. The antibody formed against all these bacterial antigens have protective function.

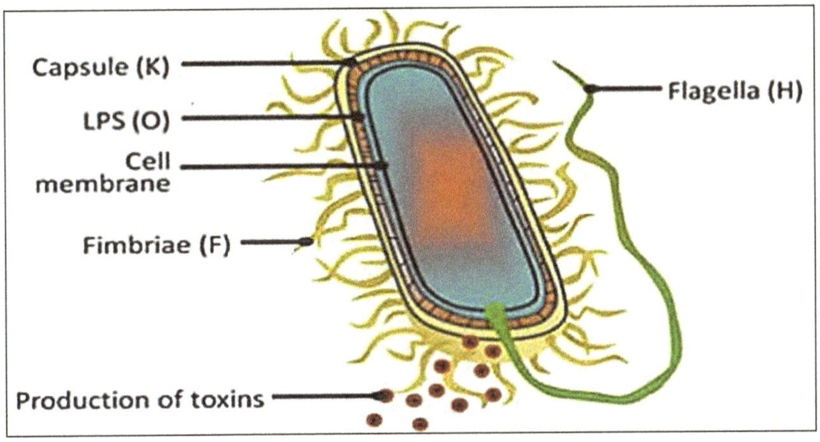

Viral Antigens

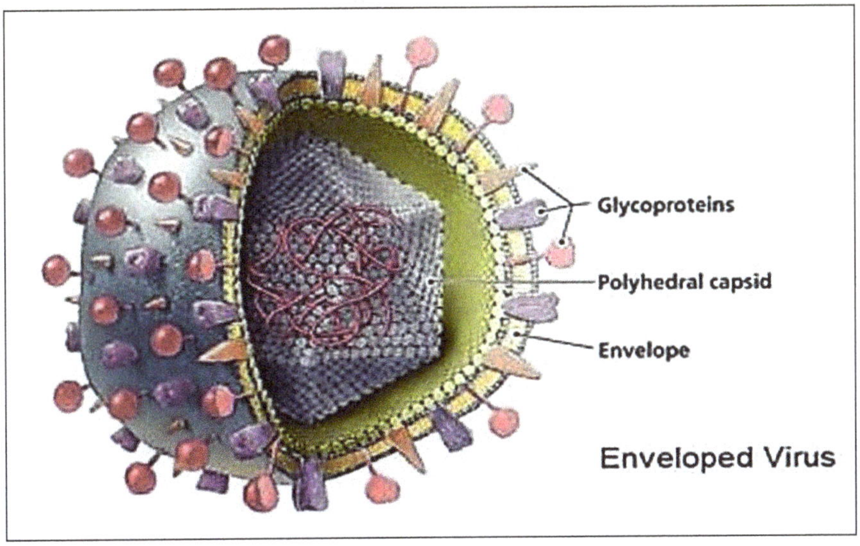

- The protein coat is called capsid.
- Capsids are good antigens.

Cell Surface Antigens of Mammalian Cells

Blood Group Antigens: Early unsuccessful blood transfusions between unrelated individuals led to the realization about A and B antigens.

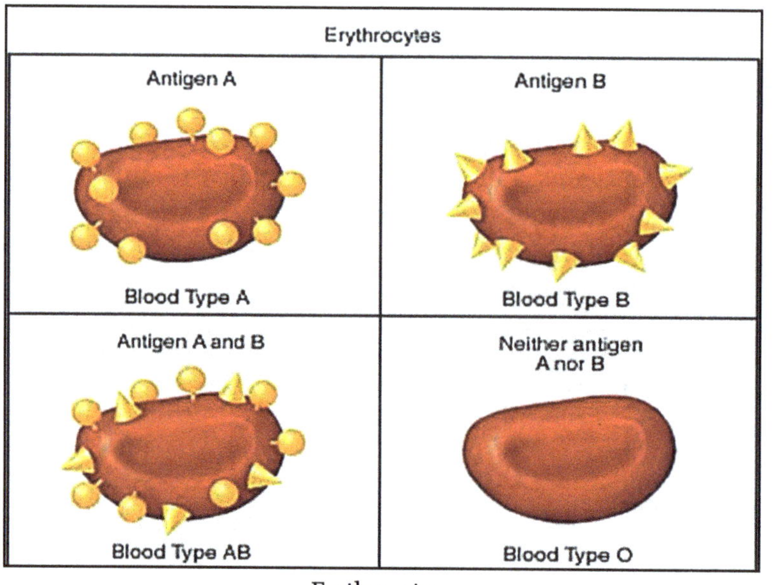

Erythrocytes.

Histocompatibility Antigens

Nucleated cells have many cell surface antigens. Most important ones are the histocompatibility antigens which provoke strong immune response and are major histocompatibility antigens.

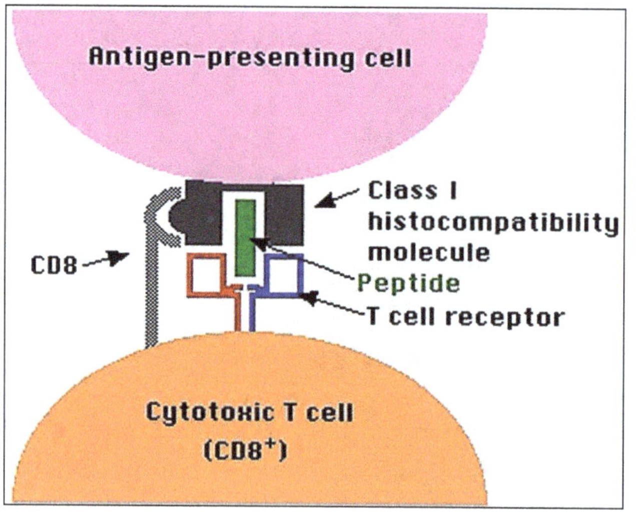

Auto Antigens and Super Antigens

Auto Antigens

Sometimes antibodies are produced in response to normal self-antigens. This happens when mechanisms responsible for self-tolerance are bypassed.

Super Antigens

It has been found later on that enterotoxins (secreted by some bacteria like S. aureus, S.pyrogens,

Pseudomonas sp.) and retroviral proteins can act as super antigens that is, provoke an exceptionally high T-cell response.

Some antigens are commonly used in experimental immunology as these are relatively cheap and available in pure form and large quantities.

Commonly used Antigens

- Endotoxins from E. coli,
- Bovine serum albumin,
- Flagellin,
- Hemocyanin: Blood pigment of keyhole limpet,
- IgG,
- Lysozyme,
- Sheep RBCs,
- TGAL: A synthetic antigen consisting of Tyr and Glu linked to Ala side chains on a Lys backbone.

Exogenous Antigens

Exogenous antigens are antigens that have entered the body from the outside, for example by inhalation, ingestion, or injection. Exogenous antigens are the most common kinds of antigens, and include pollen or foods that may cause allergies, as well as the molecular components of bacteria and other pathogens that could cause an infection.

Endogenous Antigens

Endogenous antigens are that have been generated within previously-normal cells as a result of normal cell metabolism or because of viral or intracellular bacterial infection (which both change cells from the inside in order to reproduce). The fragments are then presented on the surface of the infected cells in the complex with MHC class I molecules.

Tumor Antigens (Neoantigens)

These antigens are presented by MHC I or MHC II molecules on the surface of tumor cells.These antigens result from a tumor-specific mutation during malignant transformation of normal cells into cancer cells. Despite expressing this antigen, many tumors have developed ways to evade antigen recognition and immune system killing.

Native Antigens

A native antigen is an antigen that is not yet processed by an APC to smaller parts. T cells cannot bind native antigens, but require that they be digested and processed by APCs, whereas B cells can be activated by native ones without prior processing.

Complete Antigens and Haptens

Haptens are molecules that create an immune response when attached to proteins. Antigens are basic molecules that induce an immune response when detected by immune system cells. Antigens may be either complete or incomplete based on the nuances of their molecule structure.

Haptens

A hapten is essentially an incomplete antigen. These small molecules can elicit an immune response only when attached to a large carrier such as a protein; the carrier typically does not illicit an immune response by itself. Many hapten carriers are normal molecules that circulate through the body. When haptens and carriers combine, the resulting molecule is called an adduct, the combination of two or more molecules. Haptens cannot independently bind to MHC complexes, so they cannot be presented to T cells.

The first haptens used were aniline and its carboxyl derivatives (o-, m-, and p-aminobenzoic acid). One well-known hapten is urushiol, the toxin found in poison ivy and a common cause of cell-mediated contact dermatitis. When absorbed through the skin from a poison ivy plant, urushiol undergoes oxidation in the skin cells to generate the actual hapten, a reactive molecule called a quinone, which then reacts with skin proteins to form hapten adducts. Usually, the first exposure causes only sensitization, in which there is a proliferation of helper and cytotoxic T cells. After a second exposure, the proliferated T cells can become activated, generating an immune reaction and producing the characteristic blisters of poison ivy exposure.

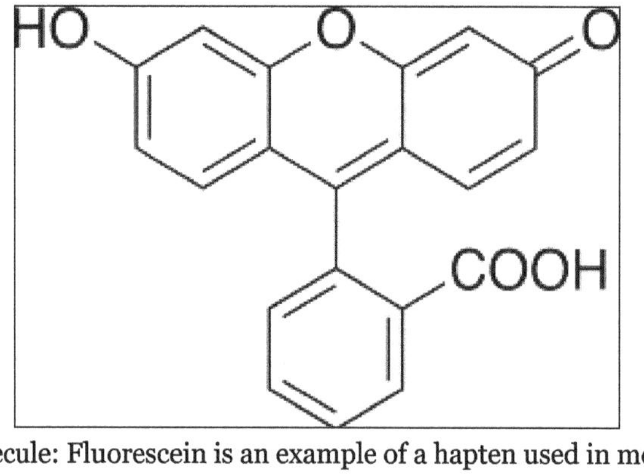

Fluorescein Molecule: Fluorescein is an example of a hapten used in molecular biology.

Some haptens induce autoimmune disease. An example is hydralazine, a blood pressure-lowering drug that occasionally causes lupus erythematosus (an autoimmune inflammatory disorder) in certain individuals with genetic predispositions to the disease. This also appears to be the mechanism by which the anesthetic gas halothane can cause life-threatening hepatitis and penicillin-class drugs cause autoimmune hemolytic anemia. Other haptens, such as flourescein, detect proteins with which they form adducts. This makes them a common part of molecular biology lab techniques.

Complete Antigens

A complete antigen is essentially a hapten-carrier adduct. Once the body has generated antibodies to a hapten-carrier adduct, the small-molecule hapten may also be able to bind to the antibody, but will usually not initiate an immune response. In most cases this can only be elicited by theonly the hapten-carrier adduct. Sometimes the small-molecule hapten can block immune response to the complete antigen by preventing the adduct from binding to the antibody, a process called hapten inhibition. In this case, the hapten acts as the epitope for the antigen, which binds to the antibodies without causing a response. If this happens with enough haptens, there will not be enough antibodies left to bind to the complete antigen, thus inhibiting the antibody response.

MHC Proteins

A major histocompatibility complex (MHC) is present on all nucleated cells of bodies of mammals. This gene complex expresses antigens (cell surface proteins) which are called MHC antigens. The discovery and the understanding of the role(s) of MHC have a long and instructive history. Its evolution is deeply rooted in the concepts of population genetics and diversity of specie. MHC plays roles in grafting of tissues and organs. But that role is incidental, its role is central to the functioning of the immune response.

Human Leukocyte Antigen (HLA)

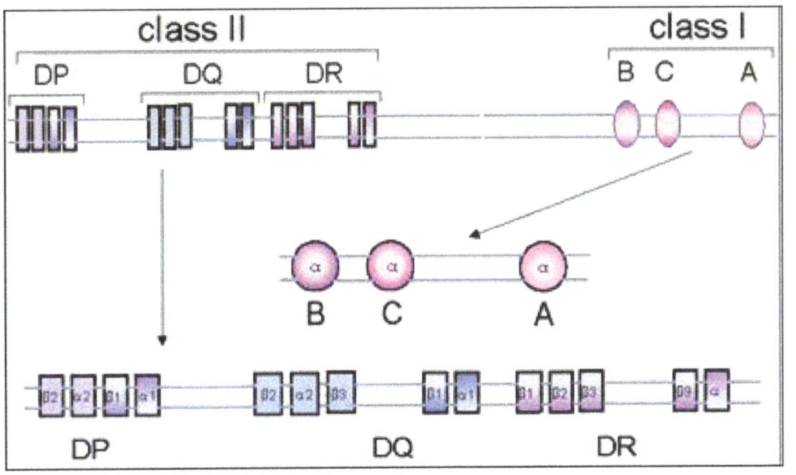

The MHC complex locus in humans is called Human Leukocyte Antigen (HLA). The corresponding loci in mice and rats are H-2 complex and RT-1 respectively. In mice, these are called H-2 genes.

The name HLA is derived from the fact that earlier it was believed that MHC antigens are present on the leukocyte only. Their genetic diversity was carried out by serological typing. In recent years, variations in MHC are investigated by genotyping. In current (unfortunate) usage MHC refers to both the genes or the cell surface proteins (antigens) and one has to rely upon the context to understand which one is being referred to.

MHC genes are most polymorphic genes known so far, that is, within a population, large variants of each gene exist. Secondly, large number of MHC Class I and MHC Class II genes exist.

Major Histocompatibility Complex (MHC)

Apart from the receptors present on the B-cells (immunoglobulins) and TCR, MHC proteins are the most diverse class of proteins. It was found out that during the graft rejection, the recipient's immune system responds to the MHC proteins of the donor. Unrelated individuals express different set of MHC proteins.

While the B-cell receptor and TCR differ from cell to cell (unless both cells are from the same clone); MHC proteins differ between individuals. Hence, MHC proteins are sometimes called markers of identity. Decades after the role of MHC proteins in graft rejection was discovered, people wondered what their role in vivo is as transfer of tissues between individuals is extremely rare in nature. Their role in directing the T-cells to the infected cells came to be discovered while many groups were trying to characterize TCR. We will revert to that when we discuss the discovery of the nature of the TCR.

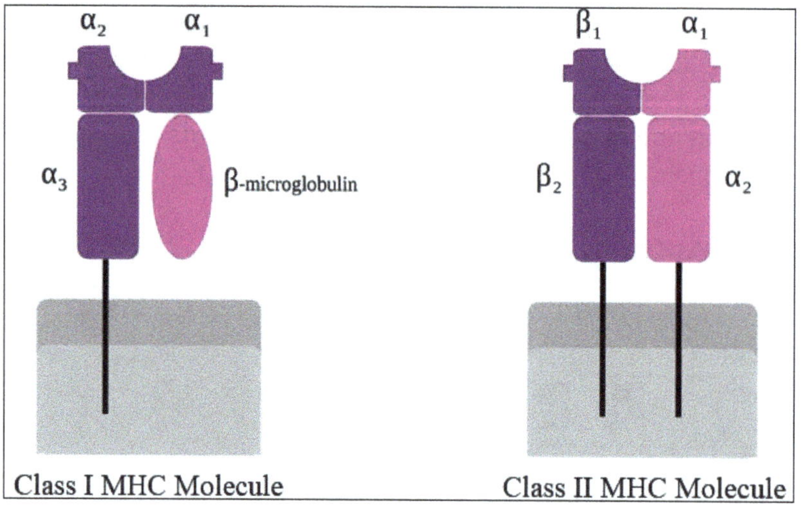

It turned out that the MHC molecules bind to the antigen derived peptides from the pathogen on the cell surface. As we have mentioned earlier, this role of MHC proteins is critical t and hence pathogens in order to survive try to down regulate the expression of the MHC protein.

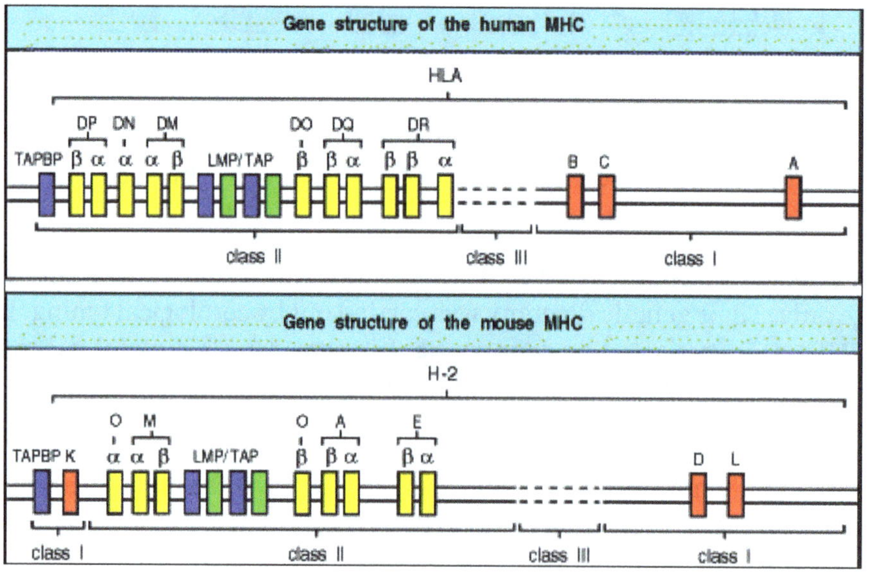

Let us look at the complete picture now. The organization of the principal MHC genes is shown for both humans (where the MHC is called HLA and is on chromo-some 6) and mice (in which the MHC is called H-2 and is on chromosome 17). The organization of the MHC genes is similar in both species. There are separate clusters of MHC class I genes (shown in red) and MHC class II genes (shown in yellow). In humans, it may span about 4 x 106 base pairs.

Tissue	MHC class I	MHC class II
Lymphoid tissues		
T cells	+++	+*
B cells	+++	+++
Macrophages	+++	++
Dendritic cells	+++	+++
Epithelial cells of the thymus	+	+++
Other nucleated cells		
Neutrophils	+++	-
Hepatocytes	+	-
Kidney	+	-
Brain	+	-†
Non-nucleated cells		
Red blood cells	-	-

All nucleated cells express MHC Class I molecules. The level of their constitutive expression varies between cell types. Cells of the immune system have high expression levels: liver cells (hepatocytes), brain cells and kidney cells have low expression levels.

RBC in mammals have practically no MHC I proteins. Thus Tc cells cannot destroy any infection in RBC. However, RBC cannot support viral replication, so that does not hurt. Malarial parasite Plasmodium, thereby, however finds RBC an immunologically previledged site. MHC class II proteins are expressed on a subset hematopoietic cells and thymic stromal cells. Thus, these are formed on B-lymphocytes, dendritic cells and macrophages.

The expression of both class I and class II proteins is regulated by cytokines (especialy interferons), IFN-γ increases the expression of both classes. It also induces the expression of MHC class II on some cell types (which do not normally express these proteins) during inflammation. Unlike humans in which activated T-cells express class II proteins, in mice T-cells do not express these proteins. In brain, except in microglia, cells do not express class II proteins.

MHC Class I Proteins

These molecules are integral membrane proteins. Each has one polypeptide chain that spans the plasma membrane. The extracellular portion has 3 domains (α1, α2, α3). The α-chain is a 44-kDa in size and is a glycoprotein. Each of the 3 extracellular or external domains contains about 90 amino acids.

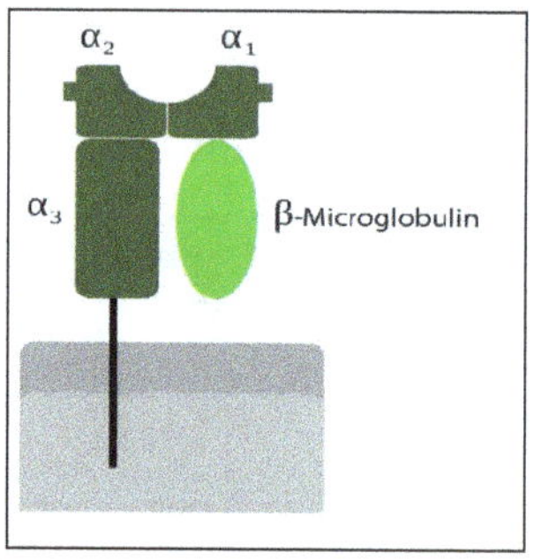

The α3 domain is linked to a domain of about 40 amino acid long transmembrane region which is followed by the cytoplasmic region of 30 amino acid. α1 domain is noncovalently linked to a single polypeptide β2-microglobulin of about 11.5 kDa. β2-microglobulin gene is on a separate chromosome number 15 on humans and 2 in mice. β2-microglobulin is critical for transport of the MHC Class I proteins to the cell surface and helps it to retain its native conformation. It is very similar to α3 domain in its amino acid composition. In fact α3 domain and β2-microglobulin have amino acid composition (and hence fold structures) similar to the C-domain in Ig.

MHC Class I Proteins: Peptide Binding Cleft

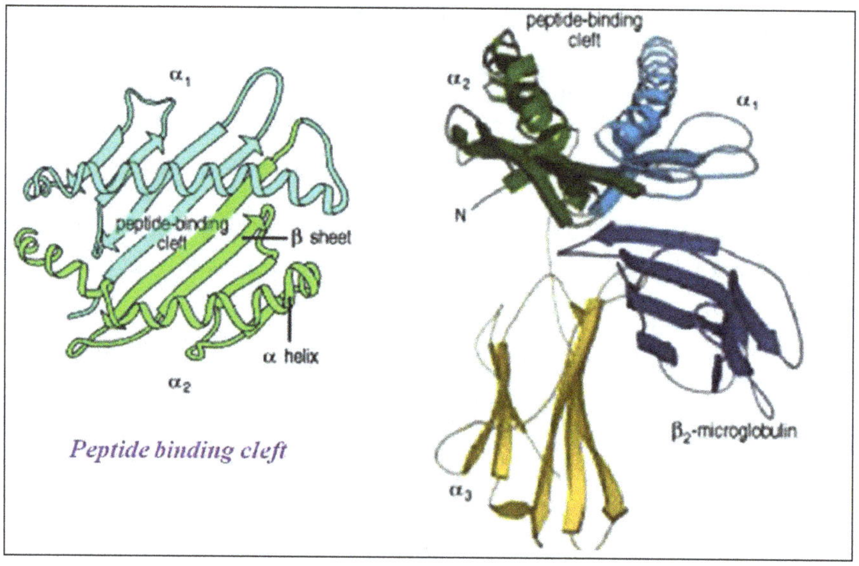

Peptide binding cleft

α1 and α2 domains fold to form two segmented α-helices resting on eight stranded antiparallel beta sheet. α1 and α2 domains form a cleft which is the site which MHC Class I uses to bind the antigenic peptide fragment. Hence, this is the site wherein polymorphism is seen.

Interferons are produced in the early phase of the viral infections as part of the innate immune response. Once again, we can see the linkage between innate immune response and adaptive immune response. Interferons IFN-α, β or γ can increase the transcription of MHC Class I α-chain and β-2 microglobulin. The human genome project has made the sequence of the HLA complex and reading frames available.

MHC Class II Proteins

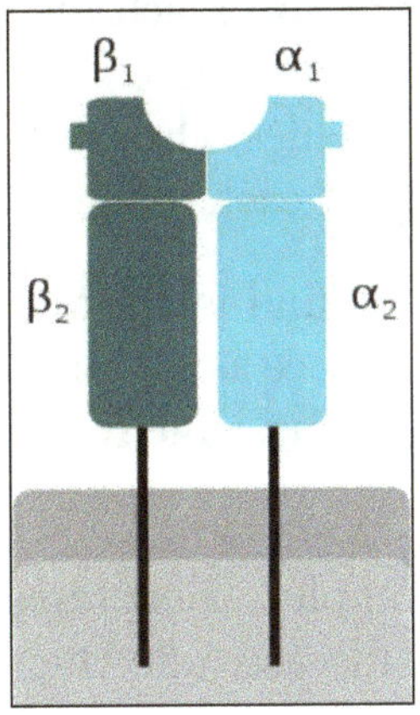

MHC Class II molecules are expressed on B-cells, macrophages, monocytes, APCs and some T-cells. These resemble Class I molecules in many respects. They consist of two transmembrane glycoprotein chains α and β of 34 kDa and 29 kDa respectively. Each chain consists of two domains: α1, α2 and β1, β2. The two chians associate noncovalently to form 4 domain appearance which is very similar to MHC Class I + β-2 microglobulin complex.

Here the bottom of the peptide binding cleft consists of eight anti-parallel beta strands and its walls have anti-parallel alpha helices. In MHC Class II, the domains part of the cleft is formed by different chains. The clefts in both MHC Class I and Class II are thus very similar. That is understandable as both do the same function: bind antigenic peptide fragments.

Detailed Map of Human MHC

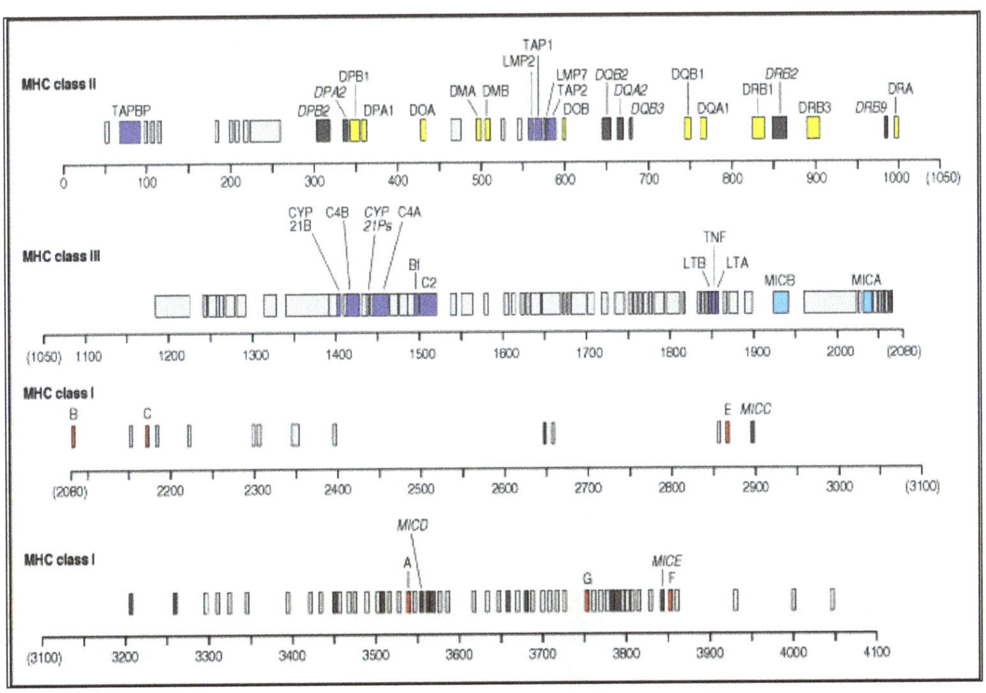

Sequences of the cloned MHC Class I and Class II genes show that each domain is encoded by a separate axon. In both mice and humans, Class I and II genes have 5' leader sequence which allows these to enter the endoplasmic reticulum and reach the cell surfaces. In addition to the highly polymorphic Class I and Class II MHC genes, there are MHC Class IB genes. These encode β-2 microglobulin associated cell surface molecules. These cell surface molecules vary in amount from cell to cell and are distributed differently on each tissue.

One such molecule of the MHC Class IB kind is H2-M3 and it can present peptides which are formylated at the N-end. This is of significance as bacterial protein synthesis starts with N-formyl-methionine. So, these are able to present antigenic fragments (from bacterial pathogens) which arise from the N-terminals.

HFe gene, also part of the same gene complex, is expressed on intestinal tract cells. Humans defective in Hfe gene have an iron storage disease called hemochromatosis. In this disease liver and other organs retain high amount of iron. Mice lacking β-2 microglobulin have similar problems.

Other genes which are part of the MHC gene complex include those which code for some complement components, some cytokines like TNF-α and TNF-β. These are sometimes called MHC Class III genes. These also include those coding for HSP7 and adrenal steroid 21-hydroxylase.

MIC gene family belongs to Class IB genes. Their prtoducts MIC A and MIC B are recognized by NK cell rceptors. This allows killing of MIC expressing cells as a part of the innate immunity even when interferons are not yet produced. These class IB genes are expressed in fibroblasts and epithelial cells especially intestinal epithelial cells.

The need for highly polymorphic nature of the MHC Class I and Class II genes is obvious. At the level of the protein products, several mechanisms exist which increases their diversity. Both polygeny and polymorphism contribute to the diversity within an individual and also in the population.

There are >200 alleles (the variant genes occupying a gene locus within a specie) of some class I and class II genes. Each allele has a high frequency in the population. Hence, there is a strong probability that both homologous chromosomes on an individual will have different allele.

Human MHC Genes are Highly Polymorphic

Expression of MHC Alleles is Codominant

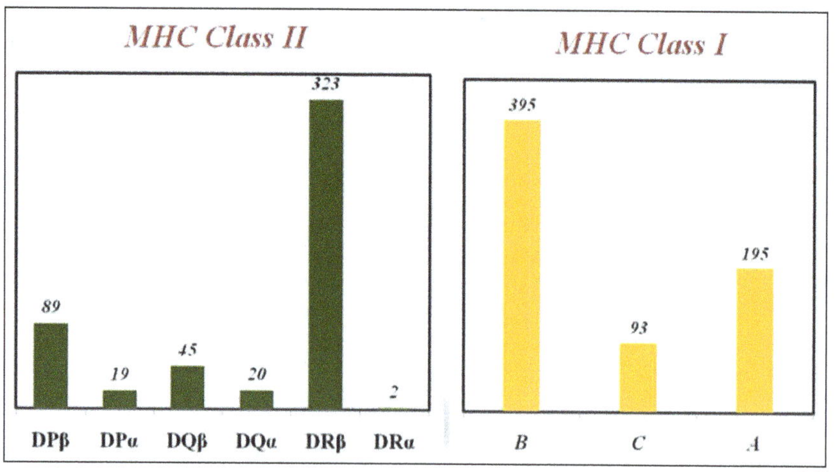

This heterozygous nature at MHC loci is the source of diversity (of protein products) as the expression of MHC alleles is codominant. Both gene products are available to present antigen fragments. This makes the number of different MHC molecules on an individual 2 times. This is in addition to the polygeny.

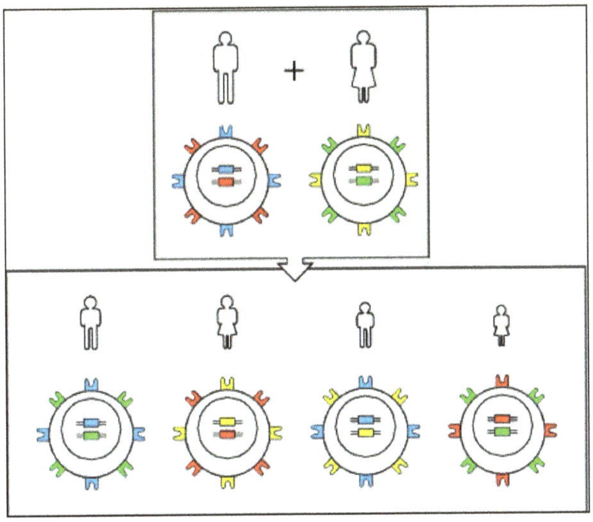

Polymorphism and Polygeny

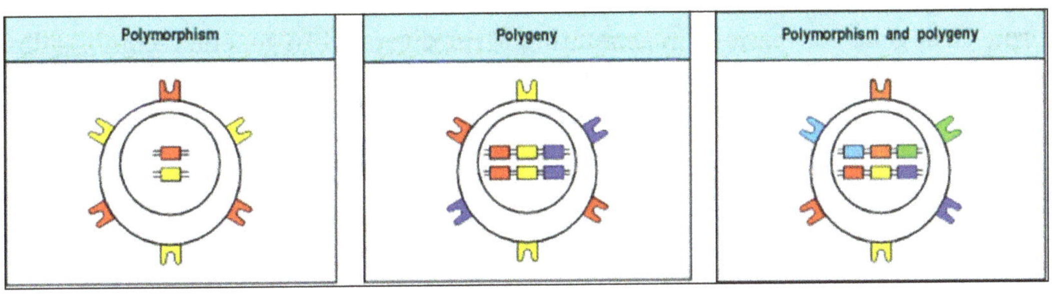

Polygeny is the availability of different related genes with similar functions. In humans, three MHC Class I genes and four sets of MHC Class II genes are available on chromosome 6. For MHC Class II genes, the diversity is increased further by combination of α and β chains encoded by different chromosomes. The diversity is more in the segments which are responsible for binding the antigenic peptide fragments.

MHC Polymorphism

Several genetic mechanisms contribute to the generation of new alleles. Apart from point mutations, gene conversion in which one sequences partly replaced by another generates MHC polymorphism.

That the selective evolutionary pressure favours MHC polymorphism is seen by the fact that point mutations are more frequently replacement mutations rather than silent mutations (change in codon but leaving the amino acid unchanged). Gene conversions result in change in the several amino acids simultaneously. As there is a close linkage between MHC genes, gene conversion is not rare in the MHC allele evolution.

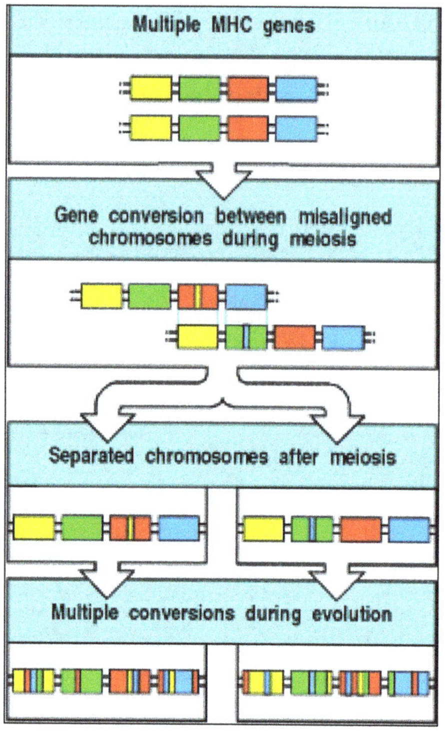

It appears that genetic recombination between different alleles at a locus may play even more important role than gene conversions in generating MHC diversity. In genetic recombination, DNA segments are exchanged between alleles on the two homologous chromosomees. Analysis of MHC sequences indicated that this has occurred frequently during the evolution. Closely related mice strains show one or two segments having been exchanged. Unrelated strains indicated accumulated effects of recombinations.

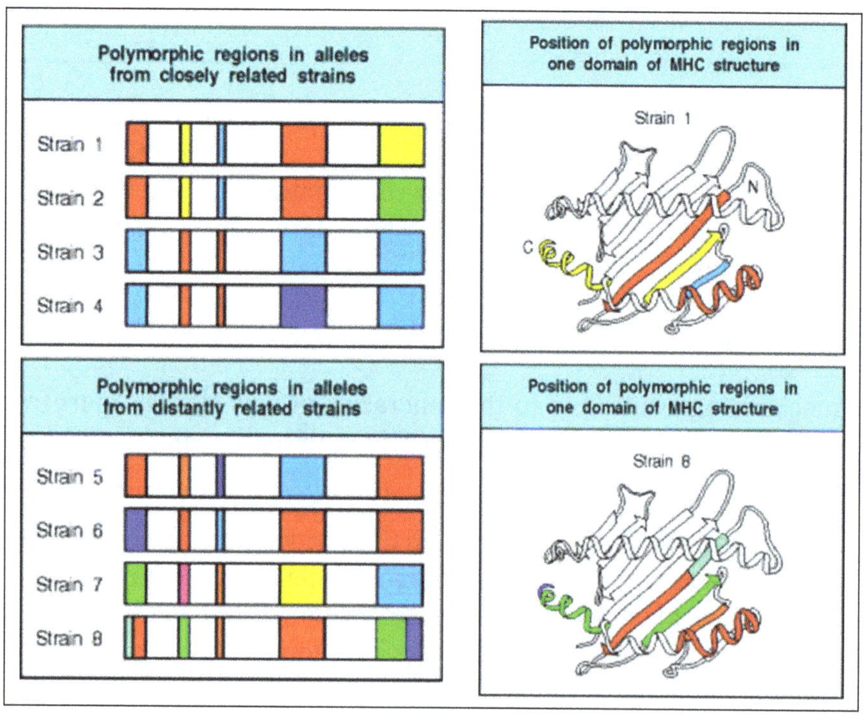

MHC Class I Like Molecules

Some MHC Class like molecules is coded by genes outside the MHC complex. CD1 is expressed on dendritic cells, monocytes and tymocytes. In antigen processing, it behaves more like MHC Class II proteins.

Apart from protein antigens, CD1 is also capable of presentation of glycolipids which are components of the myobacterial membranes. These include mycolic acid, glucose monomycolate, phosphoinositol mannosides and lipoarabinomannans. CD1 gene may have evolved for presentation of microbial lipids.

MHC Class I proteins bind to peptides from cytosolic degradation whereas MHC Class II prefer peptides generated in endocytic vescicles. Wide variety of infectious agents is one reason behind the evolutionary selective pressure for maintaining MHC diversity.

The role of MHC in Thymic Education

T-cells recognize antigenic peptide fragments only in combination with MHC molecules. The diversity of MHC molecules thus widens the range of peptides that can be presented to T-cells. T-cell responses are said to be MHC-restricted. The 'non-self' peptide must be present along with a 'self' marker: either MHC Class I or Class II antigen.

Not only that, in an individual, these have to be its own MHC proteins. These are the ones, T-cells during their development, encountered during thymic education to be positively selected. Non-self MHC cannot do this function. Hence, T-cell responses are actually self-MHC restricted.

MHC and Transplantation

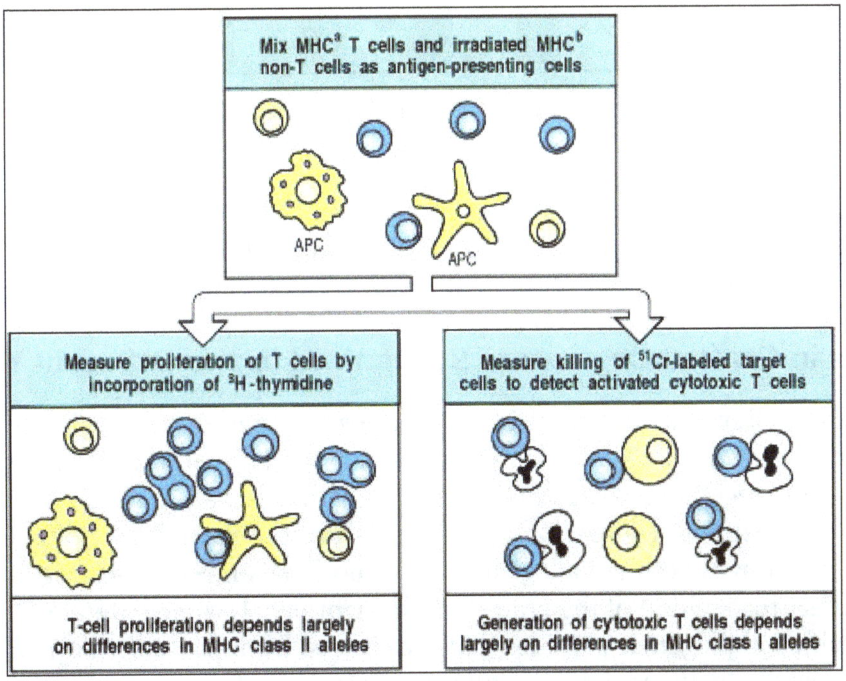

Tissue graft rejection is T-cell mediated immune rejection. MHC matching between donor and recipient increases the chance of graft acceptance. The broad guidelines for tissue grafting emerge from studies with skin grafting.

- Grafting from another organ/site of the same individual is called autograft.
- Graft from a genetically identical individual is called syngeneic graft.
- Graft from an unrelated individual is called allogeneic or allograft.

Tissue typing is carried out by a specific antiserum to cells. Complement is added. If the cell had antigen corresponding to the antibodies in the antisera, cells are lysed.

In mixed lymphocyte culture/reaction (MLC/MLR), lymphocytes from prospective donor and recipient are mixed. If the cells do not divide, they have a shared MHC specificity.

Tissues Commonly Transplanted in Clinical Medicine

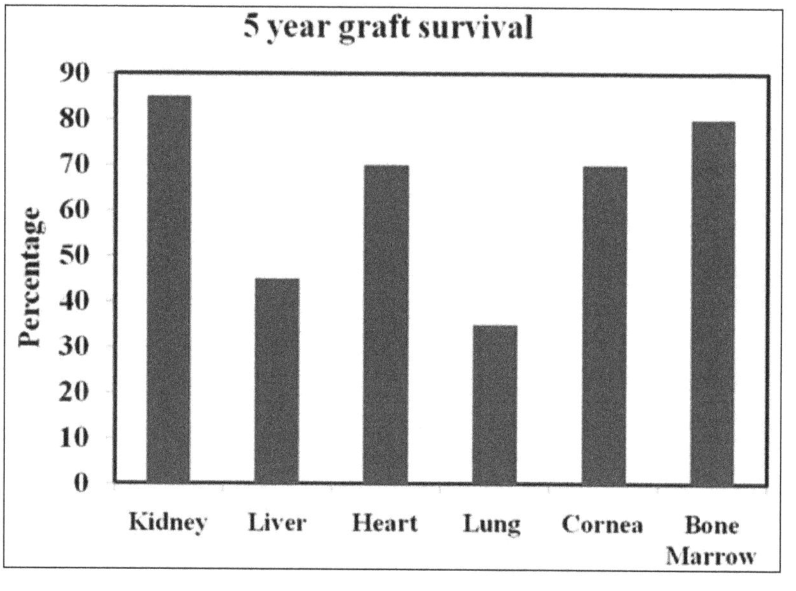

Currently it is possible to graft many organs. During grafting (except in the case of corneal graft and some bone marrow grafts), immunosupppressive drugs have to be used. Kidney and heart transplants are the most common, liver and lungs are less common but not infrequent.

Alloantigens in Grafted Organs are Recognized in two Different Ways

Both antibody and cell mediated mechanisms are known to operate but T-cells are involved in all cases except blood transfusion. Direct recognition of a grafted organ (red in upper panel) is mediated by T cells whose receptors have specificity for the allogeneic. MHC class I or class II molecule in combination with peptide. These alloreactive T cells are stimulated by donor antigenpresenting cells (APC), which express both the allogeneic MHC molecule and costimulatory activity (bottom left panel). Indirect recognition of the graft (bottom right panel) is mediated by T cells whose receptors are specific for allogeneic peptides that are derived from the grafted organ. Proteins from the graft are processed by the recipient's antigen-presenting cells and are therefore presented by self (recipient) MHC class I or class II molecules.

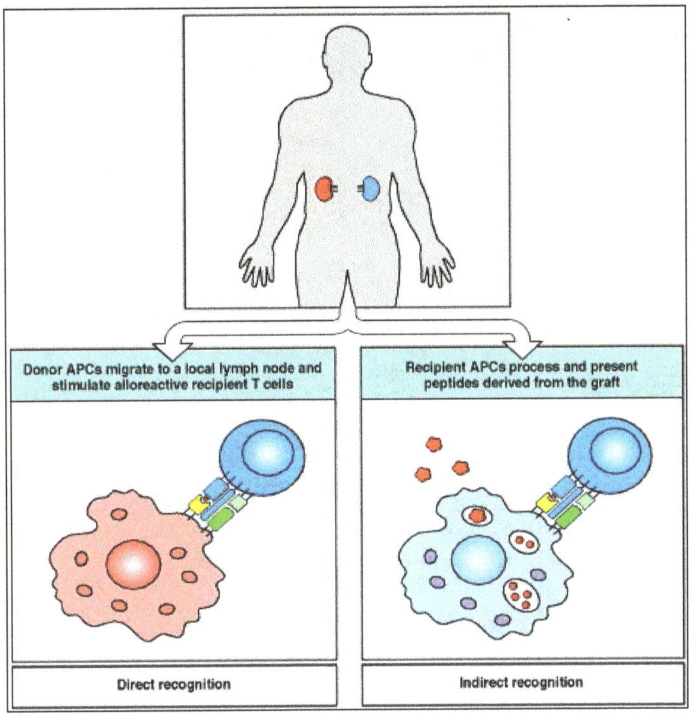

Types of Allograft Rejection

Minor H antigens are peptides derived from polymorphic proteins of the donor's cell which are different from the corresponding ones on the recipient's cell. Work with kidney transplants has led to classifying allograft rejection into three types:

- Hyperacute rejection: This occurs during time period which ranges from few minutes to few hours. Preexisting circulatory antibodies are involved and cause type II hypersensitivity. Prior blood transfusion/organ transplant definitely favours this.
- Acute rejection: This takes weeks to months and is T-cell mediated. Delayed type hypersensitivity and cytokines from monocytes/macrophages are involved.
- Chronic rejection: This takes place after months/years. It is also cell mediated and may occur due to differences in minor transplantation antigens.

Mechanism of Allograft Rejection

The % of involvement of direct and indirect recognition mechanisms is not always known. Acute rejection is largely due to direct recognition. In indirect recognition, T-cells may activate macrophages leading to tissue damage and fibrosis. This leads to antibody mediated graft rejection. The direct recognition is best understood in terms of skin grafts. Here langerhans cells act as APC. These as APCs migrate to lymph nodes where they interact with recirculating naïve T-cells specific for graft antigens and activate them. The effector T-cells reach blood via thoracic duct and from blood to the grafted tissue. As only graft is destroyed, it is mediated by adaptive immune response and not by just innate immunity processes like inflammation.

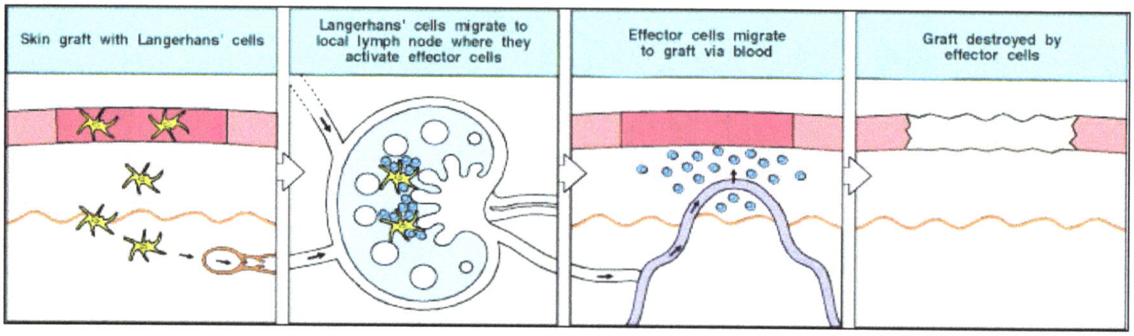

Xenotransplants

Animals have been considered as organ/tissue donors. Pigs have many organs of sizes similar to humans and have been studied widely. Such transplants from different species are called xenotransplants. Unfortunately, human shave natural haemagglutinins which bind to the cell surface carbohydrates of pigs. This leads to hyperacute rejection.

MHC molecules on pig cells would be another problem. Over and above these problems, ethical and social issues are worrisome factors. So is the possibility of unknown viruses being transferred and become infected in the germ line.

The discoveries of the MHC molecules during organ/tissue transplantation and later their involvement in T-cell responses shows how sometime two different lines of investigations can merge and help understanding biological phenomenon. Finally, MHC restriction of T-cell responses completes the circle. Foetus is a "graft" which is tolerated and not rejected under normal circumstances. That puzzling phenomenon involves both "grafting" and immunological tolerance.

Antigen Processing

T-cell recognition of antigen presenting cells depends on their expression of a spectrum of peptides bound to Major Histocompatibility Complex class I (MHC-I) and class II (MHC-II) molecules. MHC Class II molecules are present on all APCs (dendritic cells, macrophages and Bcells). Cytokines regulate the expression of MHC class II molecules on APCs.

In MHC class II processing pathway (also called exogenous pathway), pathogens are broken down by proteases into antigenic fragments. These fragments associate with MHC class II molecules which are produced in the endoplasmic reticulum.

These complexes of antigenic peptides: MHC class II molecules reach cell surface to be able to interact with $CD4^{+}T_{h}$ cells. Gut flora and food antigens do not provoke immune response under normal circumstances. Severe infections with pathogenic organisms result in overcoming competition with the gut flora.

Antigen processing is the degradation of antigen to create smaller fragments. This is followed by their association with MHC or MHC like molecules. The complex MHC: antigen peptide or a similar complex clonally selects the T-cells specifically. If the degradation is blocked, presentation cannot take place. T-cell response in that case is not available. Different cells process antigens with varied efficiency and thus are able to stimulate T-cells to a varying level.

Broadly, B-cells use Ig receptor to bind antigen via clonal selection. The specific antigen is internalized and returned to the cell surface. There the processed antigen gets associated with MHC class II molecules (present on all B-cells). The B-cell now acts as the antigen presentation cell for T_h2 cells.

Phagocytosis by mononuclear phagocytes opsonizes particles of immune complexes via their Fc and C3 receptors. Processing and presentation to T_h1 cells follows.

Immature dendritic cells phagocytocyse antigens using multiple receptors: Fc, C3, scavenger and lectin family. These dendritic cells after this migrate to lymph nodes and processed antigens are presented to T-cells there. T-cell activation requires three signals for activation:

- Antigenic peptide along with a MHC molecule to form nonself and self-molecules complex.
- Co-stimulatory signals received via CD28.
- Specific cytokines.

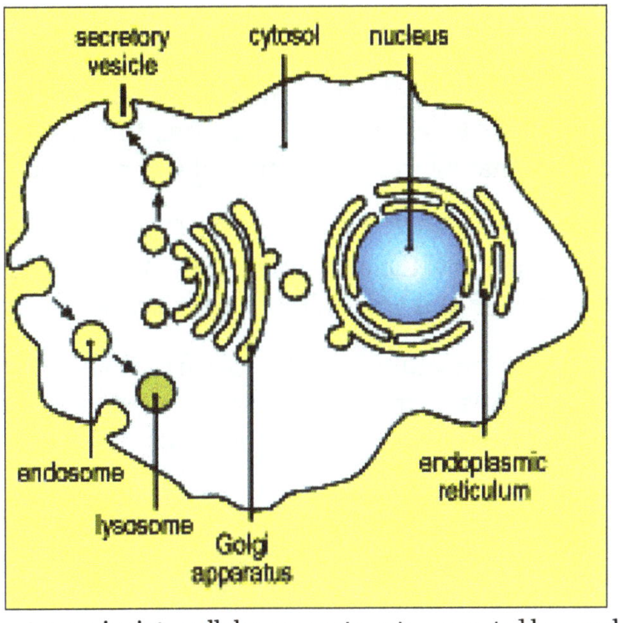

There are two major intracellular compartments, separated by membranes.

The peptides are derived from pathogens which can be present in two parts of the cell. The cell can be viewed as consisting of cytosol which is "linked" to the nucleus via pores on the membrane of the nucleus.

The vesicular system, on the other hand, is "linked" to the extracellular fluids. This includes endoplasmic reticulum, golgi bodies, endosomes, lysosomes and some other vesicles. Endoplasmic reticulum can generate vesicles which fuse with golgi membranes. The fused vesicle can reach outside the cell. This is how the movement of molecules/materials takes place from inside to outside.

In the reverse direction, cells endocytocyse molecules/particles which end up in endosomes. From there, these go into lysosome wherein hydrolases break these down. Phagocytic cells are the ones which are involved in the uptake of particles/pathogens which end up in their endosomes.

The choice of APC is made depending upon the first encounter between the pathogen and the immune system. In lymph nodes and spleen, interdigitating dendritic cells (IDCs) are the most important APCs for CD4 T–cells. IDCs, B-cells, macrophages all can express high level of MHC II molecule. However, IDCs for some reasons are more effective APCs.

For many polysaccharides, marginal zone macrophages act as APCs. This presentation can last for months or years. An antigen displayed on recirculating macrophages in medulla lasts only for days/weeks.

Viruses and some bacteria infect cytoplasm or the nuclear compartment. On the other hand, most of the pathogenic bacteria and eukaryotic parasites proliferate in endosomes and move into lysosomes. Cytosolic pathogens are taken care of by Tc (CD8) cells. The pathogens infecting other compartments are dealt with by CD4 T-cells.

Microbial antigens (include leprosy and tuberculosis causing mycobacteria) are taken up in macrophages and proliferate there. There is another route by which pathogens or their products end up in intracellular vesicles. Bacteria secreting toxins etc along with their products enter cells by phagocytosis, endocytosis or macropinocytosis.

Theses internalization mechanisms are operated by APCs which include dendritic cells, macrophages and B-cells. T-cells are MHC restricted. At the same time, there is a need to distinguish between antigenic peptides originating from cytosolic and vesicular compartments.

This is neatly achieved by involving MHC molecules. MHC class I molecules deal with cytosolic origin. In turn CD8 T cells (Tc cells) kill the presenting APC. MHC class II are associated with peptides sourced from vesicular system and are dealt with CD4 Tcells.

Macropinocytosis is carried out by immature dencritic cells in the infected tissues. These cells also use receptor mediated phagocytosis. Arising from bone marrow, these cells become localized in various tissues. Later, these cells migrate via lymph to the nearby lymph nodes to meet naïve lymphocytes.

For receptor mediated engulfment of pathogens, these cells use their receptors for common structural components of pathogens such as bacterial cell wall proteoglycans. This action is similar to that of macrophages, neutrophils. Intracellular degradation follows.

Macropinocytosis does not involve receptors; it just internalizes viruses and bacteria. Either way, the dendritic cells take up infecting pathogens while resident in tissues. Thus, activated they migrate to the lymph node and become APC for T-cells there.

Immature dendritic cells populate much of the surface eoithelia and organs such as heart and kidney. Activated dendritic cells secrete cytokines which enables these cells to infact participate in both innate and adaptive immunity.

A T-cell responds to the dendritic cell (as APC) in lymph node and become effector cell. If a T-cell runs into uninfected epithelial cell, it becomes anergic and that is a dead end for this T-cell. It no longer can become an effector cell.

The precursors of dendritic cells are myeloid cells in the bone marrow which migrate to peripheral tissues. These cells have low level of MHC molecules and do not yet have B7 costimulators. In this immature form, these cells cannot activate naïve T-cells.

Nevertheless these cells do have receptors which enable them to recognize common pathogens and thereby ingest them just like macrophages. Antigens can also be taken up by receptors such as DEC205 or nonspecifically by macropinocytosis.

When infection occurs, these immature cells get loaded with antigen, migrate to the lymph node. Once there, these lose the property of phagocytosis, start expressing B7.1 and B7.2 and high level of both MHC class I and II molecules, along with adhesion molecules ICAM-1, ICAM-2, LFA-1, LFA-3 as well as dendritic cell specific adhesion molecule DC-SIGN which can bind to ICAM-3 strongly. Important examples of such cells are Langerhan's cells of the skin. Their large granules are called Birbeck granules.

Triggered by the infection, Langerhan's cells migrate to the lymph nodes and interact with T-cells. Maturation of dendritic cells like Langerhan's cells consists of their becoming efficient APC for T-cells. Their presence of body surface complements innate immunity as part of the first parameter.

The chemokine DC-CK secreted by dendritic cells in lymphoid tissues attracts naïve T cells. Migration of immature dendritic cells to lymph nodes is a part of their life of these cells. Cells which reach end of their life span without encountering infection/antigen also travel to the lymphoid tissues. Without any co-stimulatory molecule and antigen, all these cells have are self antigens-which may be derived from any peripheral tissues. The result is tolerance.

The initiation of the recognition of the pathogen by dendritic cells involves nonclonotypic receptors. Dendritic cells play an important role in viral infections. They are ingested but not killed. Using protein synthesis capabilities of dendritic cells, viruses synthesize the protein coats. However, this does lead to the peptides from these appearing on dendritic cell surface along with MHC class I molecules.

Dendritic cells also act as APCs for bacterial and fungal pathogens. Dendritic cells also act as APCs for inhaled environmental allergens and alloantigens of transplanted organs. The stimuli activating lymphatic lining encourage migration of dendritic cells to the nearby lymph node. Hence, dendritic cells play an important role in allograft rejection. Thus dendritic cells are powerful and versatile APC. Many members of toll like receptors family are expressed by these cells. All these features widen the range of pathogens which dendritic cells can recognize.

The example of viruses infecting cells like dendrites and synthesizing proteins. That takes place in the cytosol. However, the export of the proteins to cell surface is via endoplasmic reticulum. The peptide binding to MHC class I protein take place also in the endoplasmic reticulum.

There are two ATP binding cassette (ABC) proteins which are associated with endoplasmic reticulum membrane. These are called transporter associated with antigen processing-1 and -2 (TAP-1 and TAP-2) and occur together as a sdimer. The genes for TAP-1 and TAP-2 map within MHC gene and are inducible by interferons produced in response to viral infection.

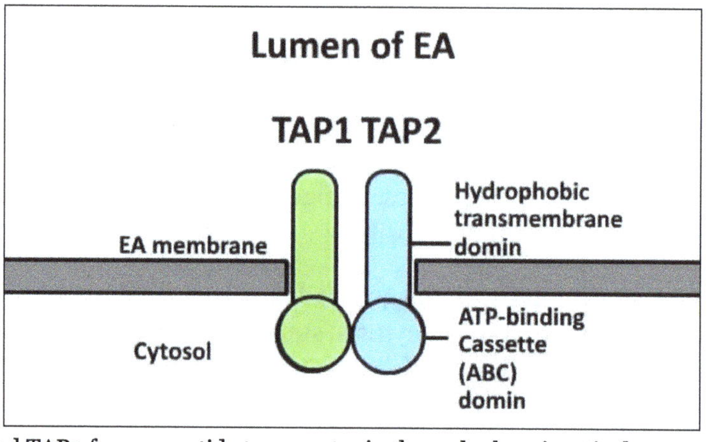

TAP1 and TAP2 form a peptide transporter in the endoplasmic reticulum membrane.

Each TAP-1 and TAP-2 has one hydrophobic domain and one ATP binding domain. There peptide transporters use energy via ATP hydrolysis. TAPs have preference for binding peptides consisting of about 8 amino acid with hydrophobic or basic amino acid at carboxyl end. MHC class I protein also have similar specificity.

Cytosolic protein degradation is carried out by proteasome mostly. A cylindrical complex of about 28 subunits present in four stacked rings (of 7 subunits each). The core constitutes the giant active site for proteolytic reaction.

Ubiquitin tagged proteins are hydrolysed by proteasome and peptides presented alongwith MHC class I molecules. Some of the proteosome's constitutive subunits are replaced by interferon inducible subunits. This changes the specificity of the proteosome to hydrolyzing bonds with hydrophobic and basic residues. The resulting peptides are preferred by MHC class I proteins for association.

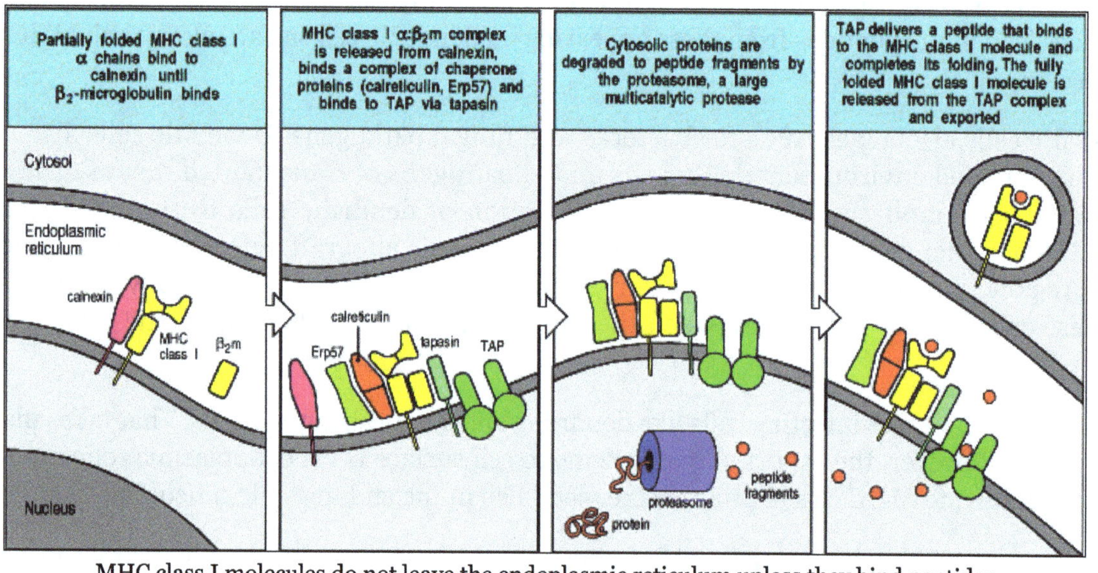

MHC class I molecules do not leave the endoplasmic reticulum unless they bind peptides.

The degraded peptides are transported into endoplasmic reticulum by TAP for binding to MHC class I molecules. The reverse-called retrograde translocation-is also possible and allows hydrolysis of misfolded proteins in endoplasmic reticulum. The membrane and secretory proteins, the viral envelop glycoproteins arising in the endoplasmic reticulum are also degraded in cytosol. The N-linked glycoprotein not only gets deglycosylated, asparagines residues also get converted to aspartic acid.

MHC class I molecules have a residence time in endoplasmic reticulum and wait there till they bind antigenic peptides. The formation of MHC molecule-peptide complex involves the binding of α-chain of the MHC molecule with β-microglobulin and this is followed by binding of the peptide. Initially, MHC molecules are in partially folded state. The formation of the complex also requires few accessory chaperon proteins.

The formation of the peptide complex releases MHC class I molecule from the endoplasmic reticulum to start its journey to the cell surfaces. In humans, α-chains of MHC class I enter the endoplasmic reticulum and bind to chaperone Calnexin. This chaperone is also responsible for association with numerous other molecules of the immune system such as T-cell receptor, MHC class II molecules and IgS.

The next step is β-microglobulin binding to the α-chain. The α: β2 microglobulin dimer dissociates in partrially folded form and binds to a complex of proteins which includes Calreticulin. A TAP-associated protein tapasin is also a part of this complex. Tapasin crosslinks TAP 1 and 2 and MHC class II. Another chaperone Erp57 which is functionally a protein disulphide isomerase participates in the peptide association of the complex. Even Erp57 and calreticulin have a general role beyond this and act as quality control molecules in assembly of many glycoproteins in endoplasmic reticulum.

The binding of peptide to the partially folded dimer releases it from the complex of TAP, tapasin, calreticulin and Erp57. This completes the process and peptide: MHC class I complex can now move to the surface through golgi. It may be noted that till the end, α:β2 microglobulin dimer remains in the partially folded state inspite of the involvement of chaperone molecules.

It is increasingly becoming clear that proteins manage multiple interactions by remaining in a disordered state which allows them to bind to multiple partners. So, the concept of what we call a folded state or native state is no longer rigid. It is likely that chaperones here are necessary to ensure that appropriate "partially folded" structure of the heterodimer is maintained before peptide is loaded.

Apart from epitope defining peptides, other peptides transported by TAP do not bind to MHC molecules in that cell and are cleared out of the endoplasmic reticulum by an ATP dependant (and TAP independent) mechanism back to cytosol. In the uninfected cells, MHC Class I molecules spent some time in endoplasmic reticulum to await infection. The ready availability of MHC class I molecules ensures that response to viral infections takes off swiftly.

Evasion mechanism of some viruses to avoid CD and T-cells is exemplified by herpes simplex virus. It prevents the transport of viral peptides by producing a TAP binding protein. On the other hand Adenoviruses produce a protein which retain MHC class I molecule in endoplasmic reticulum by binding to it.

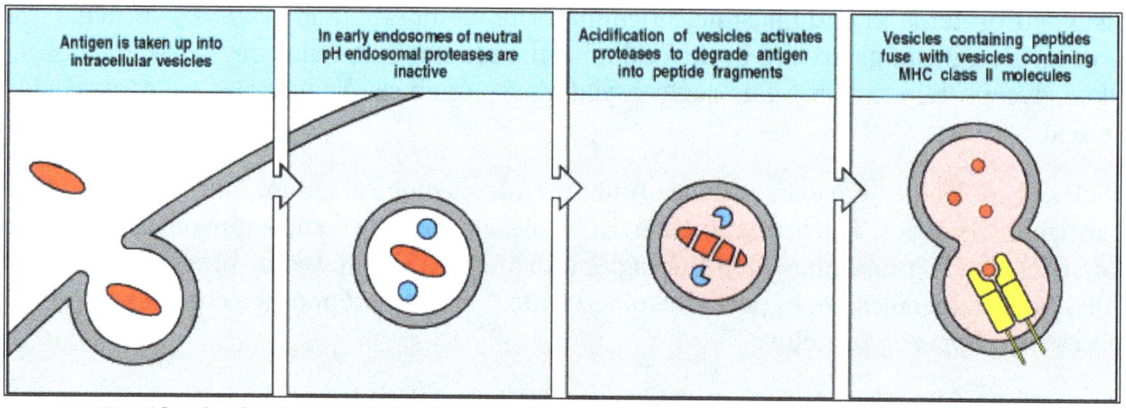

Peptides that bind to MHC class II molecules are generated in acidified endocytic vesicles.

The pathogens that replicate inside intracellular vesicles in macrophages are degraded to form peptide antigens differently. Protozoan parasite Leishmania and Mycobacteria are important examples of this class.

After macrophages activated, intravesicular proteases degrade proteins to produce peptide fragments that bind to MHC class II molecules. Extracellular pathogens and proteins internalized by endocytic vesicles follow the similar route. In both cases, CD4 T-cells recognize the peptides presented along with MHC class II molecules.

Proteins after endocytosis become part of endosomes. The endosomes gradually become increasingly acidic and finally fuse with lysosomes. The particulates entering the cell by either phagocytosis or macropinocytosis also follow this path. Some drugs such as Chloroquine can make endosome interior less acidic and prevent antigen processing/presentation. Cathepsins B, D, S, L are acid proteases, S and L seems to be more critical.

Unactivated macrophages lack both MHC class II molecules and B7 molecules. Infecting pathogens result in their expressing MHC class II and B7 molecules. Some foreign proteins which do not provoke sufficient immune response do so when mixed with some bacterial adjuvants. The latter induce the co-stimulatory activity.

This design ensures that macrophages can discharge their functions: scavenge dead or dying cells or destroying pathogens which do not require adaptive immune mechanism. For all such activities, macrophages do not need co-stimulatory activity for T-cells. Kupffer cells and macrophages of spleenic red pulp are prime scavengers for cleaning dead cells from blood every day.

B-cells as APCs

Macrophages are more efficient at taking up particulate matter as compared to soluble antigens. Immature dendritic cells, however, can take up soluble anbtigens from extracellular fluids by macropinocytosis.

B-cells complement these by specifically binding soluble antigens through the Ig receptors. We have often referred to this. Let us look at the details of what happens inside the B-cells. B-cells express high levels of MHC class II molecules constitutively. The result is high levels of antigen presentation in the form of MHC class II antigenic peptide complexes.

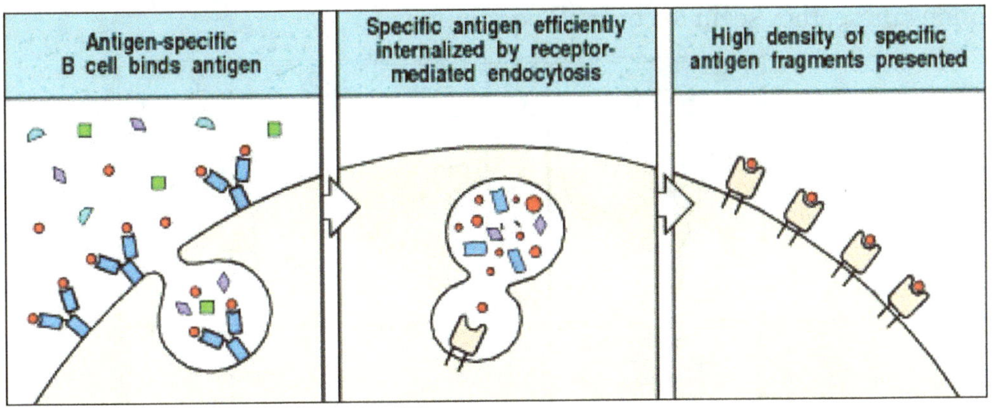

B cells can use their immunoglobulin receptor to present specific antigen very efficiently to T cells.

This in turn leads to their interaction with CD4 T-cells. This activates naïve CD4T cells who in turn drive the B-cell differentiation and response further. This T-cell interaction requires presence of co-stimulators on B-cells. Just like dendritic cells and macrophages, B-cells are induced to express co-stimulatory molecules like B7.1 and B7.2. Historically, B7.1 was identified as a molecule expressed in response to the presence of microbial polysaccharides.

Many soluble antigens such as HEW lysozyme, ovalbumin and lysozyme are able to provoke B-cell response only when bacterial adjuvants are used in vitro. This is understandable as immune system must have evolved to take care of microbial infections and not injections of soluble antigens.

B-cells (and other APCs) in the absence of costimulating signals fail to activate naïve T-cells. Instead, they turn T-cells into anergic or unresponsive state. These mechanisms seem to be additional safety features so that immune response is not initiated against self-antigens. Inside B-cells, the antigen is hydrolysed in intracellular vescicles. It is here that it binds to MHC class II molecules. These vescicles as such are transported to the cell surface.

It appears that B-cell specificity is based upon the quantitative criterion of how good has been the internalization of the antigen. Apparently, even non-specific antigens are internalized, albeit very poorly by B-cells. The corresponding antigenic fragments are also presented but the cell surface density of antigenic fragments: MHC complexes are very poor.

While dendritic cells resident in tissues lack specificity of B-cells and hence cannot concentrate these antigens, their initial encounter with the naïve T-cell specific to the antigen (derived fragment) is higher than B-cells. Once the T-cell has stayed in the lymphoid tissue upon antigen presentation by a dendritic cell, it can encounter specific B-cell there with a greater probability.

Thus the roles of various APCs: dendritic cells, macrophages and B-cells are complementary as far as activating naïve T-cells is concerned. Dendritic cells cover a wide range as far as variety of antigens is concerned. Macrophages act as APCs only for those pathogens which they cannot destroy on their own. So, they do not seek T-cell help unless necessary.

B-cells are more specific towards antigens which they can recognize via clonal selection. Only dencritic cells which are non-phagocytic lymphoid resident express costimulators constitutively. Macrophages and B-cells have inducible co-stimulatory molecules. The three classes of APCs differ in their antigen uptake mechanism, MHC class II expression and the type of antigens which they cater to. Among these, they seem to cover all possibilities.

	Dendritic cells	Macrophages	B cells

Antigen uptake	+++ Macropinocytosis and phagocytosis by tissue dendritic cells Viral infection	Phagocytosis +++	Antigen-specific receptor (Ig) ++++
MHC expression	Low on tissue dendritic cells High on dendritic cells in lymphoid tissues	Inducible by bacteria and cytokines – to +++	Constitutive Increases on activation +++ to ++++
Co-stimulator delivery	Constitutive by mature, nonphagocytic lymphoid dendritic cells ++++	Inducible – to +++	Inducible – to +++
Antigen presented	Peptides Viral antigens Allergens	Particulate antigens Intracellular and extracellular pathogens	Soluble antigens Toxins Viruses
Location	Lymphoid tissue Connective tissue Epithelia	Lymphoid tissue Connective tissue Body cavities	Lymphoid tissue Peripheral blood

The properties of the various antigen-presenting cells.

Oral Tolerance

Foreign antigens ingested as food do not provoke immune response. As these are not self antigens present in thymus and bone marrow, the corresponding lymphocytes are not eliminated that is, negatively selected for destruction.

Experiments with fed ovalbumin show that even along with adjuvants, the antibody response is veryu slow to develop. This suppression mechanism is specific as any antigen injected produces antibody response by the animal. Hence, it appears that antigens taken up by oral route involve peripheral tolerance. This is known as oral tolerance.

Before trying to understand oral tolerance, it is necessary to consider bacterial infections of the gut. We have a symbiotic bacteria (about > 400 specie), much of these in the colon ilieum. Competitively, these keep away pathogenic bacteria colonizing the gut. Some situations such as breaks in mucosal lining of the gut, these non-pathogenic strains of bacteria (E.coli is an important example) crosses over to blood and initiate systemic infection. Thus, lining of the gut is responsible for ensuring that the relationship with the gut flora remains symbiotic.

Infact animals which are guotobiotic (delivered and kept into germ free environments) have poorly developed immune responses. So, in some ways gut flora is necessary to keep adaptive immune responses to a desired level but the immune responses do not eliminate gut flora.

Nevertheless gut is often invaded by pathogenic microbes. Enteric bacteria (Salmonella, Yevsinia.

Shigella, Listeria) protozoa (Entamoeba histolytica, cryptosporidium) and many viruses infect through gut. Inflammatory response involving cytokines and chemokines induce adaptive immune responses. Dendritic cells and other APCs initiate clonal selection of the lymphocytes.

Some pathogens enter enterocytes, the cells lining intestine. This leads to signals in the form of released cytokines and chemokines: chemokines IL-8, MCP-1, MIP-1α, MIP-1β, RANTEs. Stressed enterocytes also start expression of non-classical MHC such as MIC-A and MIC-B which bind to γ- δ T-cells. These kill infected mucosal cells and this initiates repair and recovery of the mucosa.

Other pathogens such as polio, retroviruses are transported by M-cells and are able to establish systemic infection. Enteric bacteria (S.typhii and S.typhimurium) also follow this route and a bacterium reaches lymphoid tissues. In the absence of the inflammatory response, the foreign antigens seem to induce tolerance in mucosal immune system. So, while pathogen cause inflammation, food antigens do not

Multiple hypothesis have been given to further explain oral tolerance. The γ- δ T-cells which have extensive presence in the mucosa are probably involved in developing oral tolerance. Some regulatory T-cells such as T_h3 and T_R1 also seem to play a role. These cells seem to suppress antigen specific responses when challenges with antigens occur.

Antibodies

Antibody is also called immunoglobulin, a protective protein produced by the immune system in response to the presence of a foreign substance, called an antigen. Antibodies recognize and latch onto antigens in order to remove them from the body. A wide range of substances are regarded by the body as antigens, including disease-causing organisms and toxic materials such as insect venom.

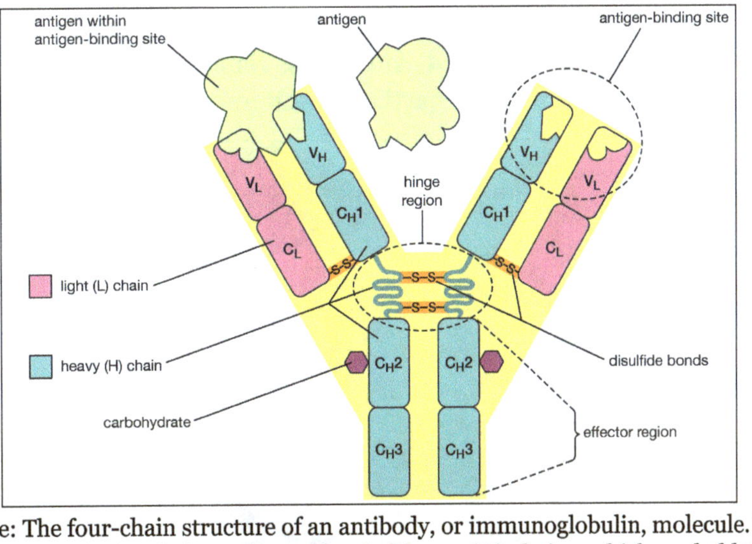

Antibody structure: The four-chain structure of an antibody, or immunoglobulin, molecule. The basic unit is composed of two identical light (L) chains and two identical heavy (H) chains, which are held together by disulfide bonds to form a flexible Y shape. Each chain is composed of a variable (V) region and a constant (C) region.

When an alien substance enters the body, the immune system is able to recognize it as foreign because molecules on the surface of the antigen differ from those found in the body. To eliminate the invader, the immune system calls on a number of mechanisms, including one of the most important—antibody production. Antibodies are produced by specialized white blood cells called B lymphocytes (or B cells). When an antigen binds to the B-cell surface, it stimulates the B cell to divide and mature into a group of identical cells called a clone. The mature B cells, called plasma cells, secrete millions of antibodies into the bloodstream and lymphatic system.

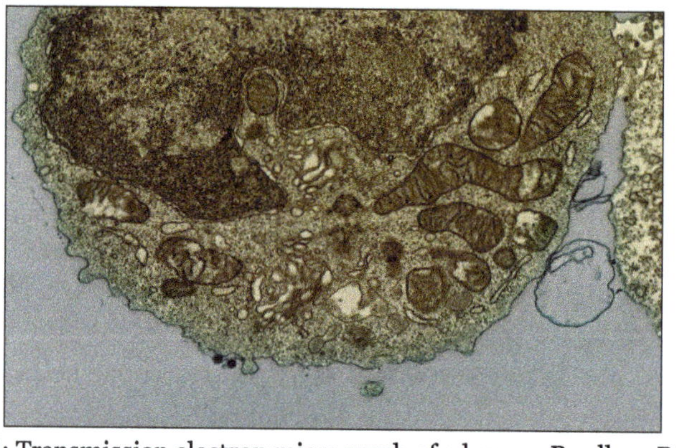

Human B cell: Transmission electron micrograph of a human B cell, or B lymphocyte.

As antibodies circulate, they attack and neutralize antigens that are identical to the one that triggered the immune response. Antibodies attack antigens by binding to them. The binding of an antibody to a toxin, for example, can neutralize the poison simply by changing its chemical composition; such antibodies are called antitoxins. By attaching themselves to some invading microbes, other antibodies can render such microorganisms immobile or prevent them from penetrating body cells. In other cases the antibody-coated antigen is subject to a chemical chain reaction with complement, which is a series of proteins found in the blood. The complement reaction either can trigger the lysis (bursting) of the invading microbe or can attract microbe-killing scavenger cells that ingest, or phagocytose, the invader. Once begun, antibody production continues for several days until all antigen molecules are removed. Antibodies remain in circulation for several months, providing extended immunity against that particular antigen.

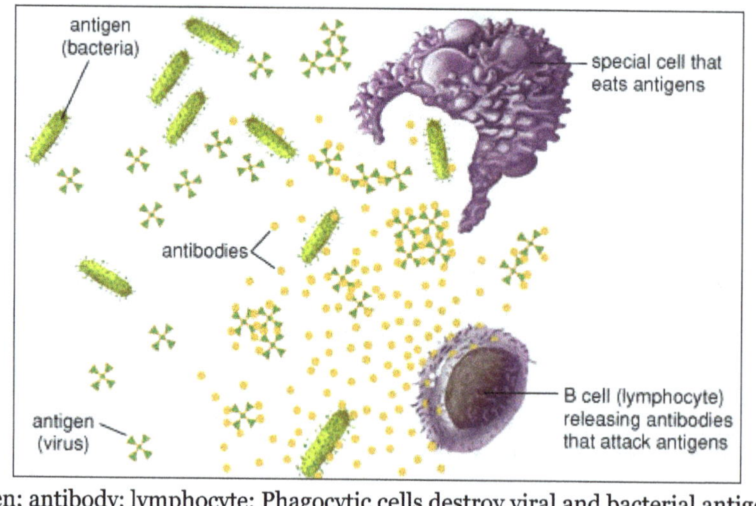

Antigen; antibody; lymphocyte: Phagocytic cells destroy viral and bacterial antigens by eating them, while B cells produce antibodies that bind to and inactivate antigens.

B cells and antibodies together provide one of the most important functions of immunity, which is to recognize an invading antigen and to produce a tremendous number of protective proteins that scour the body to remove all traces of that antigen. Collectively B cells recognize an almost limitless number of antigens; however, individually each B cell can bind to only one type of antigen. B cells distinguish antigens through proteins, called antigen receptors, found on their surfaces. An antigen receptor is basically an antibody protein that is not secreted but is anchored to the B-cell membrane. All antigen receptors found on a particular B cell are identical, but receptors located on other B cells differ. Although their general structure is similar, the variation lies in the area that interacts with the antigen—the antigen-binding, or antibody-combining, site. This structural variation among antigen-binding sites allows different B cells to recognize different antigens. The antigen receptor does not actually recognize the entire antigen; instead it binds to only a portion of the antigen's surface, an area called the antigenic determinant or epitope. Binding between the receptor and epitope occurs only if their structures are complementary. If they are, epitope and receptor fit together like two pieces of a puzzle, an event that is necessary to activate B-cell production of antibodies.

Each antibody molecule is essentially identical to the antigen receptor of the B cell that produced it. The basic structure of these proteins consists of two pairs of polypeptide chains (lengths of amino acids linked by peptide bonds) that form a flexible Y shape. The stem of the Y consists of one end of each of two identical heavy chains, while each arm is composed of the remaining portion of a heavy chain plus a smaller protein called the light chain. The two light chains also are identical. Within particular classes of antibodies the stem and the bottom of the arms are fairly similar and thus are called the constant region. The tips of the arms, however, are highly variable in sequence. It is these tips that bind antigen. Thus each antibody has two identical antigen-binding sites, one at the end of each arm, and the antigen-binding sites vary greatly among antibodies.

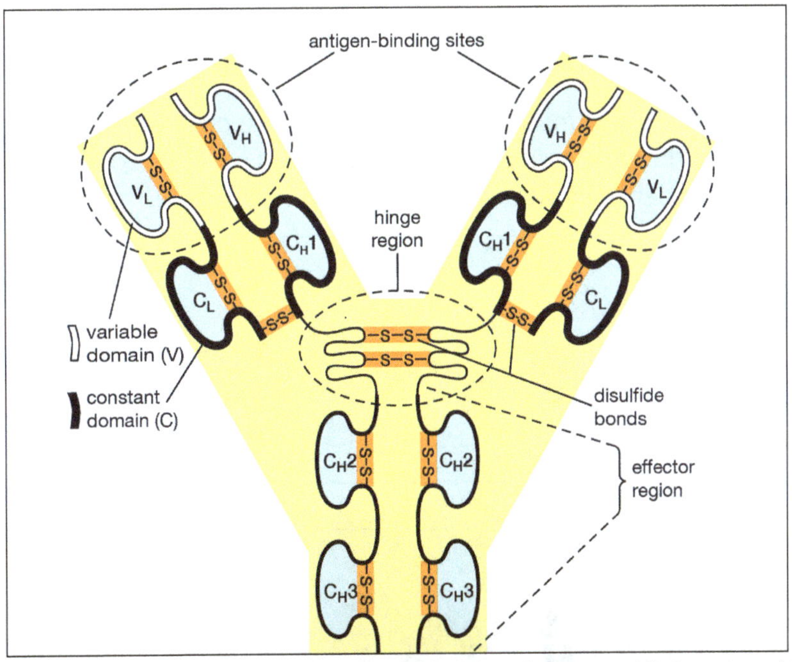

Variable and constant domains of an antibody: Variable (V) and constant (C) domains within the light (L) and heavy (H) chains of an antibody, or immunoglobulin, molecule. The folded shapes of the domains are maintained by disulfide bonds.

Antibodies are grouped into five classes according to their constant region. Each class is designated by a letter attached to an abbreviation of the word *immunoglobulin*: IgG, IgM, IgA, IgD, and IgE. The classes of antibody differ not only in their constant region but also in activity. For example, IgG, the most common antibody, is present mostly in the blood and tissue fluids, while IgA is found in the mucous membranes lining the respiratory and gastrointestinal tracts.

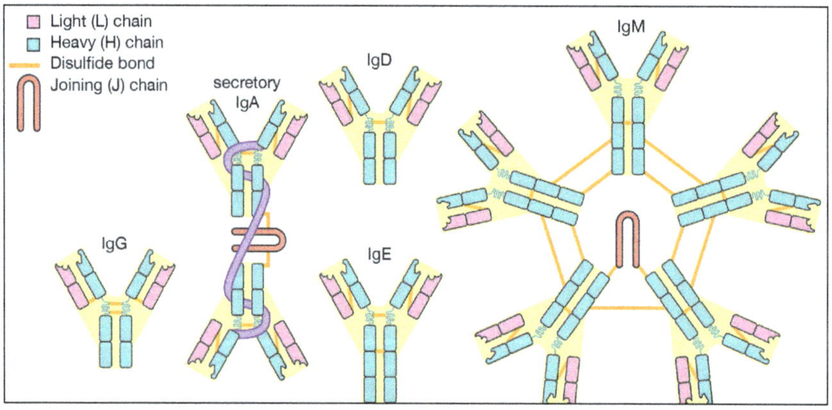

Classes of antibodies: The five main classes of antibodies (immunoglobulins): IgG, IgA, IgD, IgE, and IgM.

Preformed antibodies, which are derived from the blood serum of previously infected people or animals, are often administered in an antiserum to another person in order to provide immediate, passive immunization against fast-acting toxins or microbes, such as those in snakebites or tetanus infections.

Ig Classes

Antibodies are present in the serum and these proteins were originally detected there. The early electrophoretic studies of the serum revealed the presence of several proteins with their characteristic migration patterns under the influence of the electric field.

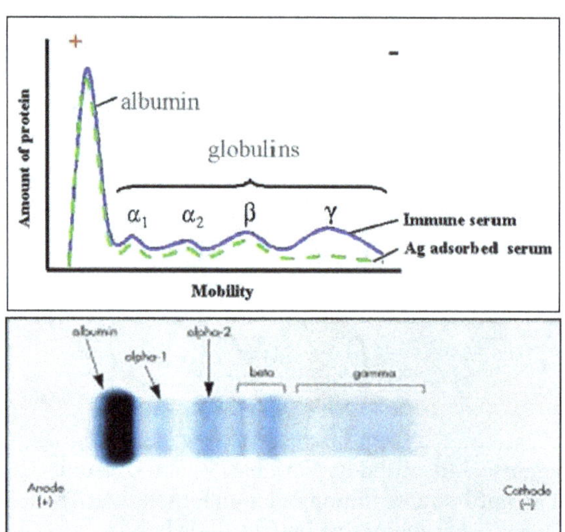

The band called γ-globulins represents immunoglobulins and it is this class which constitutes antibodies. The terms antibodies and immunoglobulins are often used interchangeably. These proteins are produced by the immune system in response to the exposure to antigens.

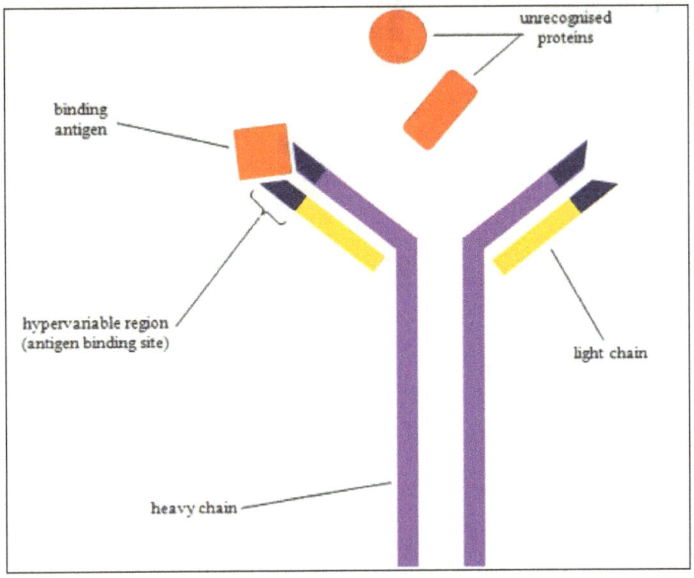

The structure of an Antibody.

As all animals are invariably exposed to antigens, some antibodies are always present in the serum. Any intentional or accidental exposure to specific antigens produces the corresponding specific antibody population in the serum.

The first breakthrough came from looking at a homogeneous protein obtained from tumours of plasma cells. All immunoglobulins are specific to that particular antigen but each in fact is specific to a particular antigenic determinant.

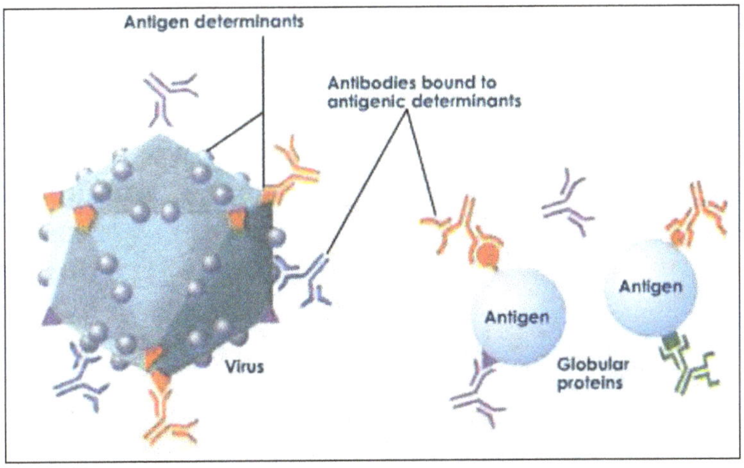

Formation of antigen-antibody complex.

For much of the work in immunology, especially in the early days, the serum as such constituted the antibody preparation. When used as such, it is, for obvious reasons, often referred to as anti-serum. In humans, there are five classes of antibodies or immunoglobulins. The different types of immunoglobulins are of different shapes.

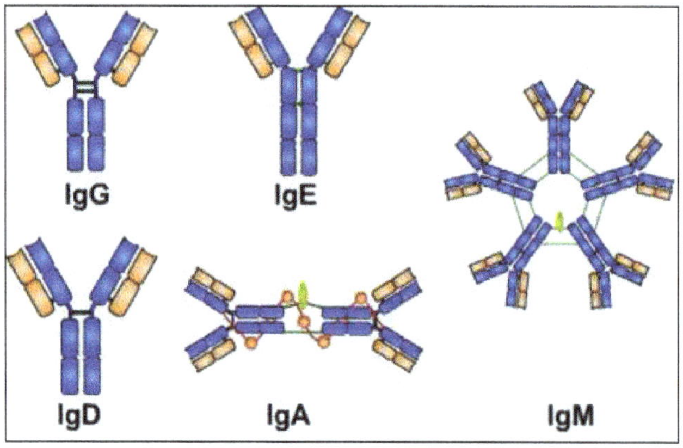

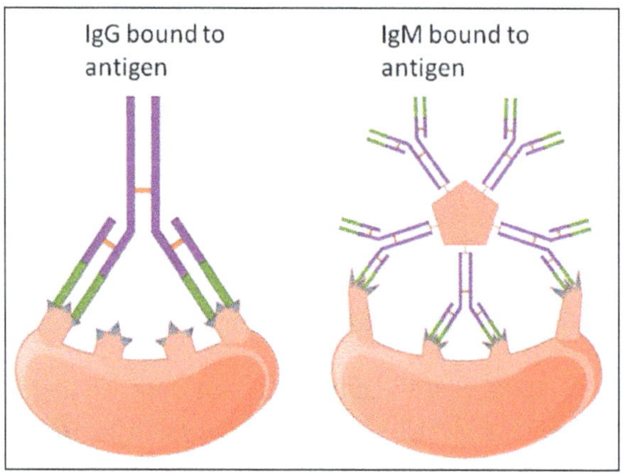

Immunoglobulin G (IgG)

This is responsible for the major part of the immunity against blood borne infectious agents. It has the highest concentration of 8-16 mg/mL in normal adult serum as compared to all other classes of the immunoglobulins. It has four different sub classes:

IgG1, IgG2, IgG3, IgG4

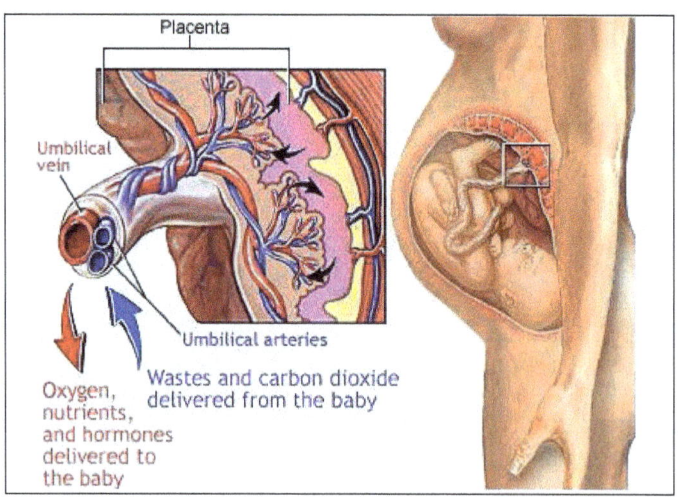

These different classes differ slightly in the amino acid sequences in a particular region called constant region in the H-chain (We will look at the different polypeptide chains of immunoglobulins shortly). These differences are reflected in their functional activities. Ig is the only immunoglobulin which can cross the placental barrier and thus provides passive humoral immunity to the developing foetus.

While all the IgG subclasses contain antibodies to components of many disease-causing bacteria and viruses, each subclass serves a slightly different function in protecting the body against infection. For example, IgG1 and IgG3 subclasses are rich in antibodies against proteins such as the toxins produced by the diphtheria and tetanus bacteria, as well as antibodies against viral proteins. In contrast, IgG2 antibodies are predominantly against the polysaccharide (complex sugar) coating (capsule) of certain disease-producing bacteria (such as, Streptococcus pneumoniae and Haemophilus influenzae).

The IgG in the bloodstream is 60-70% IgG1, 20-30% IgG2, 5-8% IgG3 and 1-3% IgG4. The amount of the different IgG subclasses present in the bloodstream varies with age. For example, IgG1 and IgG3 reach normal adult levels by 5-7 years of age while IgG2 and IgG4 levels rise more slowly, reaching adult levels at about 10 years of age. In young children, the ability to make IgG2 antibodies to the polysaccharide coatings of bacteria develops more slowly than the ability to make antibodies to proteins.

IgG subclass deficiencies affect only IgG subclasses (usually IgG2 or IgG3), with normal total IgG and IgM immunoglobulins and other components of the immune system being at normal levels. These deficiencies can affect only one subclass or involve an association of two subclasses, such as IgG2 and IgG4.

IgG2 or IgG3 deficiencies are the most common IgG subclass deficiencies. Since IgG1 comprises 60% of the total IgG level, deficiency of IgG1 usually drops the total IgG level below the normal range, resulting in hypogammaglobulinemia.

IgG4 is present in very low levels in children younger than 10 years of age, so IgG4 deficiencies are not usually diagnosed before age 10. IgG4 may be undetectable in the serum of many "normal" adult individuals, and therefore low IgG4 alone is insufficient evidence of an antibody deficiency disorder requiring Ig replacement.

All patients with IgG subclass deficiency require more extensive diagnostic evaluation including the demonstration of a poor antibody response to vaccine challenge before the patient is diagnosed with a clinically significant IgG subclass deficiency necessitating specific treatment that may include Ig replacement therapy.

IgG subclass deficiencies may be associated with other immunoglobulin abnormalities. One common pattern is IgG2 and IgG4 subclass deficiency associated with IgA deficiency. IgG subclass deficiencies are also an integral component of other well-known primary immunodeficiency diseases, such as Wiskott - Aldrich syndrome and Ataxia-Telangiectasia.

IgG subclass deficiencies are sometimes associated with poor or partial responses to pneumococcal polysaccharides, specifically IgG2 deficiency with or without IgG4 deficiency. Recently, a number of inflammatory diseases, including some forms of pancreatitis, were found to be associated with an elevated IgG4 level.

Ig Properties

IgG molecules can agglutinate or clump particulate antigens such as blood cells and microorganisms. So, soluble multivalent antigens form precipitates while insoluble antigens form clumps. In both cases, the insoluble antigen-antibody complexes become the target of phagocytic cells and are destroyed. Hence, Agglutination serves the purpose of a defence mechanism as a part of the immune system response.

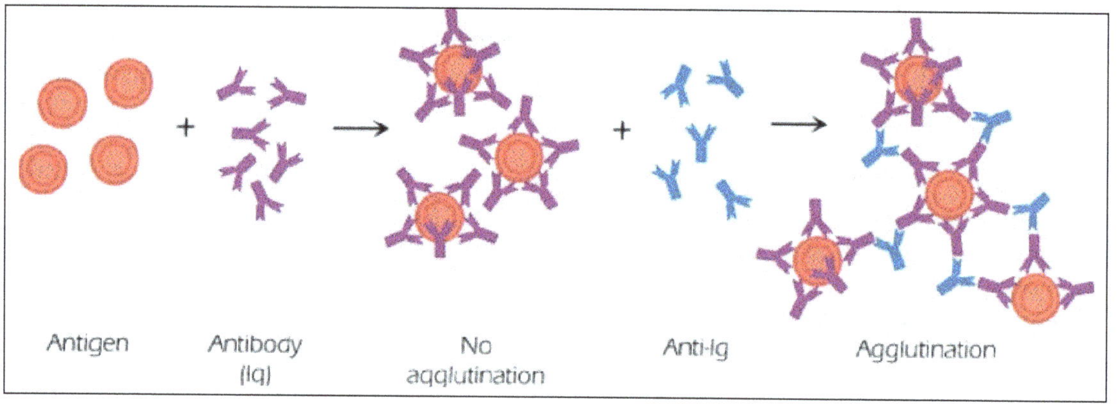

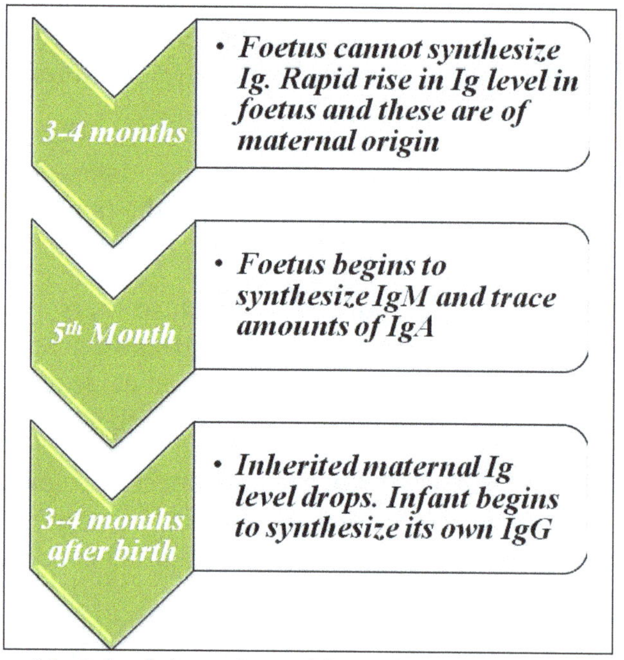

Timeline of the Ig levels in mother and foetus during pregnancy and beyond.

All IgG subisotypes except IgG2 can cross the placental barrier. IgG in fact is the only isotype which can do so. This enables the immune system of the mother to protect the growing foetus. Analysis of fetal immunoglobulins (of maternal origin since the foetus is not synthesizing antibodies) at the 3rd or 4th month of pregnancy shows a rapid rise in the level.

Hence, the foetus and the neonate is protected by the maternal IgG. F_c fragment is the enabler for placental passage as $(F_{ab})2$ or F_{ab} are unable to cross placental barrier.

Blood ABO Compatibility

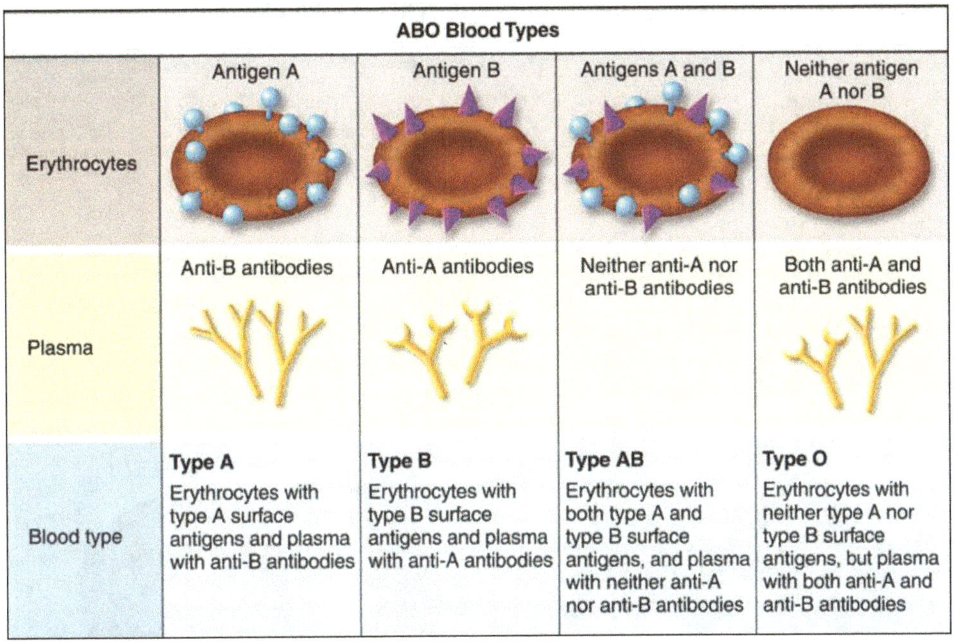

Type A blood has 'A' surface antigen on RBC and plasma has antibodies against 'B' antigens. Type B blood has 'B' surface antigen on RBC and plasma with anti A antibodies. Type AB blood has both 'A' and 'B' surface antigens on RBC and plasma does not have antibodies against either A or B. Type O blood has no surface antigens on RBC but plasma has both anti A and anti B antibodies.

Erythroblastosis Fetalis

In 1930, Landsteiner observed that RBC from about 85% of humans agglutinated when rabbit antisera raised against rhesus monkey RBC was added. This was traced to the presence of a specific antigen on the surface of RBCs of this population. RBCs of the other 15 % humans did not have this surface antigen. This human RBC antigen was named Rh antigen. Around the same-time, two obstetricians had a case where a woman gave birth to a baby with haemolytic disease of the new born (erythroblastosis fetalis). The woman reportedly earlier had a severe reaction when given a transfusion of her husband's blood with the blood type. Luckily collaboration between Landsteiner and these people ensued and it was found that while the husband had the same blood type under the ABO system, he was Rh^+. Mother's blood was Rh^-. The baby's blood was Rh^+.

Subsequent work showed that in all cases of erythroblastosis fetalis, mother's blood was Rhand baby's blood was Rh^+. This happened when father had Rh+ blood. It was further found that mother's serum in such cases has antibodies against Rh+ antigen. These antibodies were of IgG isotype. So, these IgG crossed the placental barrier and reacted with the Rh+ RBC of the foetus (which had Rh antigen) and caused haemolysis. Small amount of Rh+ RBC from the foetus can cross the placental barrier. Immune system of the mother responds to these. This is normally too low titre to cause a problem. Hence, it is generally not a problem during the first pregnancy. During separation of the placenta, release of the cord blood into the mother exposes her to a large number of Rh+ RBC from the foetus. This produces large number of antibodies against Rh antigen in mother. Hence, during second and successive pregnancies, severe immune reaction occurs.

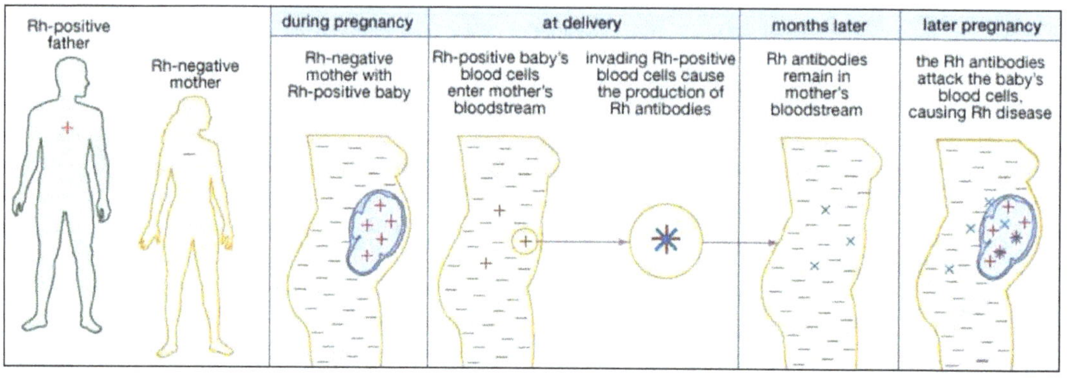

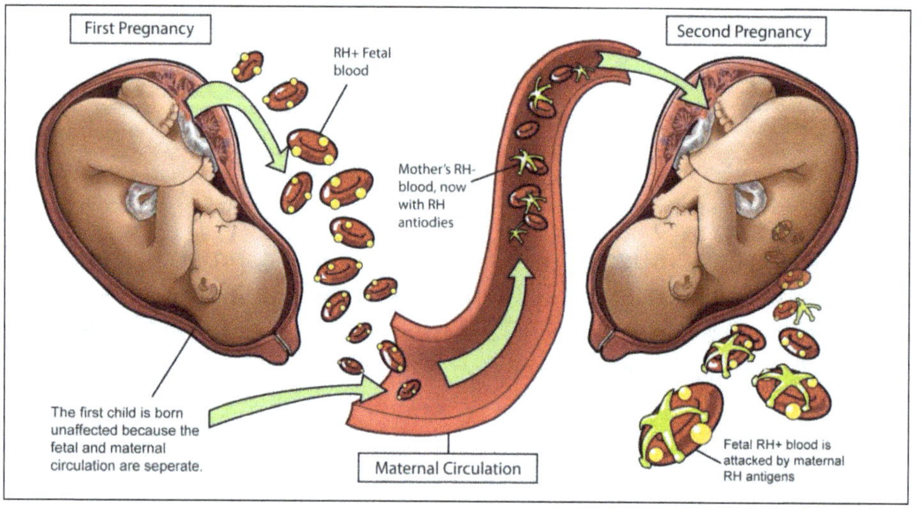

Rh antigen(s) are encoded by a single locus which has closely linked genes and many alleles. The various Rh antigens are Dce, DCE, DcE, Dce, dCe, dcE and dCE. D antigen among all Rh antigens is the strongest immunogen and majority of the cases of the haemolytic diseases of the neonate are traced to the absence of this antigen in the mother. Administration of human anti-Rh antibodies within 72 h of posturition clears up the Rh+ cells from the mother‟s blood. Human IgG against the D antigen works in most of the cases.

Blood type	Antigens present	Antibodies in serum
AB+	ABH, RH D	None
AB-	ABH	Rh D
A+	AH, Rh D	B
A-	AH	B, Rh D
B+	BH, Rh D	A
B-	BH	A, Rh D
O+	H, Rh D	AB
O-	H	AB, Rh D

Why Similar Incompatibility with ABO Blood Groups does not Cause this Problem?

In case of ABO incompatibility, immunoglobulins primarily are of IgM isotype and do not cross the placenta. IgG even if made are adsorbed by the placental tissues as these contain AB blood group substances. Rh antigen is not present in such tissues as Rh antigen is present only in RBCs.

Opsonization

Opsonin in Greek language means to prepare for eating! IgG coats the microorganism by recognizing various epitopes via its F_{ab} portion. Many phagocytes including macrophages and PMN phagocytes have receptors for Fc portion and are thus able to recognize "ready for phagocytosis" microorganisms. Whole microorganism is engulfed and destroyed by the phagocytic cell.

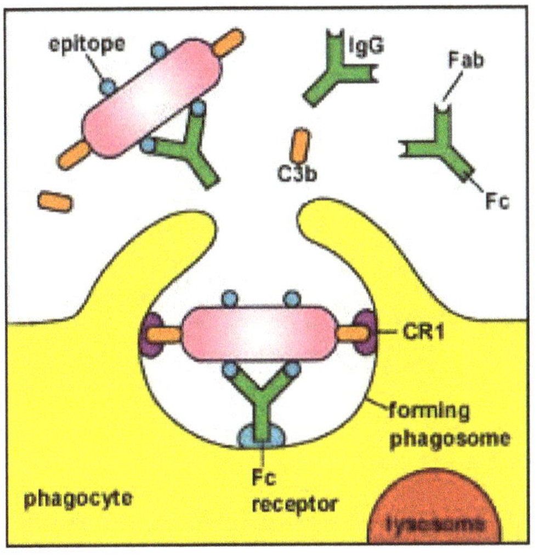

Antibody Dependant Cell Mediated Cytotoxicity (ADCC)

In this case, large granular lymphocytes, macrophages or granulocytes bind to the IgG bound microorganism or a tumour. This binding is again via their receptors for Fc portion of the antibody. The cells destroy the microbe/tumour by releasing various substances. Macrophages release enzyme and ROIs whereas LGLs use perforin and cytokines. NK cells mediated cytotoxicity operates similarly and is mediated by LGLs.

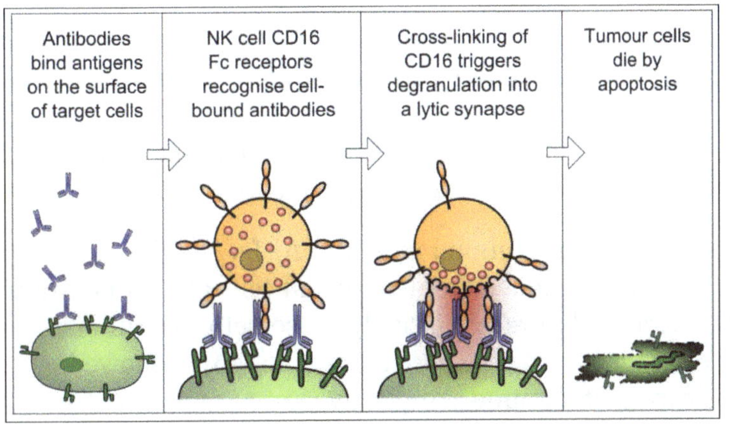

Neutralization of Toxins

IgG is the preferred isotype for passive immunization against toxins such as tetanus and botulinus.

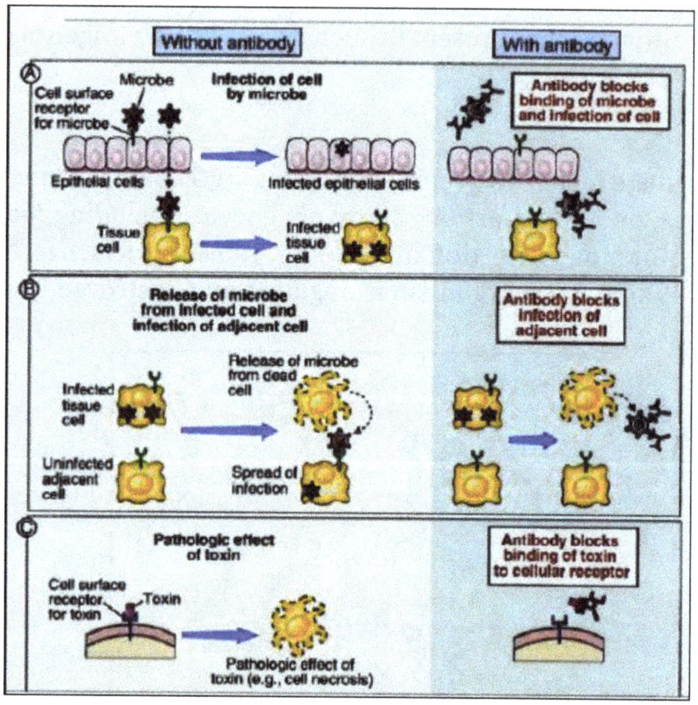

Immobilization of Bacteria

IgG specific to the flagella and cilia of mobile bacteria cause them to clump. This prevents their ability to spread in the tissue and invade the adjoining tissues.

Virus Neutralization

Both inhibition of viral attachment to the target cell (by IgG against viral coat protein) and inhibition of viral penetration or shedding of the viral coat (so that released DNA/RNA can lead to virus replication) are known.

Properties of IgA

Immunoglobulin A (IgA) is the most abundant immunoglobulin found in humans. IgA, at concentrations of about 2–4 mg/ml, is considered as the second most prevalent antibody in serum after Immunoglobulin G (IgG), which is normally present at about 12 mg/ml. IgA constitutes 10%–15% of the total immunoglobulin in serum and is the predominant immunoglobulin class found in external secretions which play key roles in immune protection.

In the resistance against infection, Immunoglobulin A (IgA) is the first line of defense. It works via inhibiting bacterial and viral adhesion to epithelial cells and by neutralization of viruses and bacterial toxins, both extra- and intracellularly. IgA also helps to eliminate pathogens or antigens via an IgA-mediated excretory pathway where binding to IgA is followed by poly immunoglobulin receptor-mediated transport of immune complexes.

IgA exists in two forms, monomeric and polymeric. Serum IgA is predominantly monomeric IgA1 which is produced in the bone marrow, while in external secretions most of the locally produced IgA is polymeric with a relative increase in the proportion of IgA2 comprising mainly dimeric forms connected by a joining peptide chain.

Secretory IgA exists in a dimer form and provides the primary defense mechanism against some local infections because of its abundance in mucosal secretions (e.g., saliva and tears). IgA appears selectively in the seromucous secretions, such as saliva, tears, nasal fluids, sweat, colostrum, milk, and secretions of the lung, genitourinary and gastrointestinal tracts, where it defends the exposed external surfaces of the body against attack by microorganisms.

This is an important function of IgA as approximately 40 mg of secretory IgA/kg body weight is transported daily through the human intestinal crypt epithelium to the mucosal surface as compared with a total daily production of IgG of 30 mg/kg. The deficiency of IgA in individuals has an increased incidence of respiratory infections associated with a lack of secretory IgA in the respiratory system.

Properties of IgM

IgM is the earliest immunoglobulin to be synthesized by the fetus beginning by about 20 weeks of age. IgM cannot cross the placenta:

- The presence of IgM in the fetus or newborn indicates intrauterine infection (The IgM must have been produced by the baby itself, as IgM cannot cross the placenta of pregnant women, placental transfer is ruled out) and its detection is useful in the diagnosis of congenital infections such as Toxoplasmosis, syphilis, rubella, cytomegalovirus infection and HIV infection.

IgM is the first immunoglobulin class produced in a primary response to an antigen. IgM antibodies are relatively short lived, disappearing earlier than IgG:

- Presence of specific IgM against a pathogen in the serum of patients indicates recent/acute infection and helps in disease diagnosis. IgM ELISA test can be performed to detect such antibodies in the serum of patients.

Pentameric IgM has 10 antigen binding sites and 5 Fc portions held together by disulfide bridge bonds and J chains:

- IgM has highest avidity of all immunoglobulins for antigens with multiple epitopes.
- Less IgM than IgG is required to neutralize viral infectivity. IgM is highly effective against viruses.
- IgM is a better agglutinin than other antibody isotypes. It takes 100 to 1000 times more molecules of IgG than of IgM to achieve the same level of agglutination.
- IgM is more efficient than IgG in activating classical pathway of complement. Complement activation requires two Fc regions in close proximity, and the pentameric nature of a single molecule of IgM fulfills this requirement.

- Monomeric IgM is the major antibody receptor on the surface of B lymphocytes for antigen recog-nition.
- Large Size: Because of its large size, IgM remains largely confined to bloodstream (80%). IgM does not diffuse well and therefore is found in very low concentration in the intercellular tissue fluids:
- IgM is believed to be responsible for protection against blood invasion by microorganisms. IgM deficiency is often associated with septicemias.
- High Molecular Weight: Because of its high molecular weight (900,000- 1000,000), it is often called macroglobulin and the 'millionaire molecule'.
- Presence of J chains: The presence of J chain allows IgM to bind to receptors on secretory cells, which transport it across epithelial linings to enter the external secretions that bathe mucosal surfaces. Although IgA is the major isotype found in these secretions, IgM plays an important accessory role as a secretory immunoglobulin.

Properties of IgD

IgD represents about 0.25% of the total serum immunoglobulins and has a M_r of 185,000 and a half-life of 2.8 days, similar to that of IgE. Its synthesis rate is at least 10 times lower than that of IgA, IgM, and IgG. The turnover rate is 37% of the intravascular pool per day, much higher than that of IgG, IgA, and IgM but less than that of IgE. The catabolism of IgD is decreased at high concentrations in serum. Initially, it was considered that about 75% of IgD was intravascular.

The IgD molecule has a long "hinge" region (between Fab and Fc) that appears to render the molecule very susceptible to proteolytic degradation with production of Fab and Fc fragments and also renders the molecule flexible, thus enhancing antigen binding. The human IgD molecule has three constant-region domains.

During storage, plasma proteases, especially plasmin, fragment the IgD molecule; the cleavage is initiated near the carboxy terminus and later in the inter-Fd-Fc area. The increased sensitivity to "spontaneous" fragmentation is considered to be also due to the Fc fragment, which is less compact for IgD than for other immunoglobulins.

IgD does not induce passive cutaneous anaphylaxis in the guinea pig; does not bind to normal lymphocytes, neutrophils, or monocytes; and does not cross the placenta. Aggregated monoclonal IgD consumed C3 to C9 without loss of the early components of the complement cascade, whereas native IgD has very little, if any, activating activity on the components of the alternative pathway of complement. IgD binds bacteria nonspecifically through the Fc fragment; for example, it binds strongly to *Neisseria catarrhalis*and*Haemophilus influenzae* and weakly to streptococcal groups A, C, and G. IgD has antibody activity to specific antigens; individuals with high levels of IgD can produce specific IgD antibodies after antigenic challenge. Among 66 selected patients, 3 had IgD antibodies to cow's milk (against bovine gamma globulin), 2 of them had antibodies to bovine serum albumin, 1 had antibodies against diphtheria toxoid, and all 4 also had antibodies of other classes and had evidence of vigorous immunologic responsiveness. Specific IgD antibodies were also described against thyroglobulin, insulin in individuals with diabetes, wheat in patients with celiac disease, tartrazine in certain patients with chemical hypersensitivity, and hepatitis B core antigen.

In 1972, IgD was discovered on the surface of many B cells; in humans and other species, IgD is the major antigen receptor isotype on the surface of most peripheral B cells, where it is coexpressed with IgM. Few plasma cells that contain only IgD could be found in lymphoid tissues; adenoids had greater numbers of IgD-containing plasma cells than did human spleen, lymph nodes, or intestinal lymphoid tissue. There is spontaneous synthesis and secretion of IgD from the tonsil lymphocytes in culture. The heavy chains of IgD from the surface of B cells are covalently linked by only one disulfide bridge, close to the carboxy terminus. This allows a higher degree of freedom of the antigen-binding sites than for the other immunoglobulin isotypes. Serum IgD and membrane-bound IgD are antigenically similar, but they differ in susceptibility to proteolysis by plasmin. Also, the δ chains from membrane-bound IgD have slightly slower electrophoretic mobility than serum monoclonal δ chains in sodium dodecyl sulfate-polyacrylamide gel electrophoresis, suggesting a larger molecular weight for the former.

The signals transduced by IgM or IgD receptors on B lymphocytes are the same but with different kinetics. Signals transmitted through surface IgD caused induction of up to 80 different somatic mutations. The majority of the cell-bound IgDs are of the κ type, while 90% of the monoclonal IgDs in serum are of the λ type. The so-called IgD paradox is the preferential association between δ heavy chains and λ light chains in IgD-secreting cells. This association is expressed in vitro and in vivo, is a property of normal human IgD-secreting cells, and is closely related to the secreted form of IgD. Explanations given for the preferential association are that physical constraints favor secretory δ-λ-chain association during intracellular chain assembly, IgD immune response is highly restricted, and there is a high degree of selectivity and unorthodox regulation of IgD light-chain expression. Human neoplastic B cells that expressed surface immunoglobulins of the λ type frequently exported IgD, whereas B cells that expressed immunoglobulins of the κ type did not export IgD (but inserted IgD into their surface membrane). This implies that possession of a λ chain facilitates the IgD secretory pathway.

IgD production is influenced by two major factors: an independent mechanism (only for IgD), which may be genetically determined, and also the same factors that control the production of the other immunoglobulin isotypes. The genes for the C region of μ and δ (C_μ and C_δ) are transcribed in the same RNA (a common transcription unit) and can be expressed simultaneously. IgM and IgD molecules present on individual B cells have identical V regions and antigen-binding sites and are coexpressed by differential (alternate) splicing of a common *VDJ* exon to either C_μ or C_δ region exons to produce separate μ and δ RNAs. Allelic exclusion exists for IgD, as for the other immunoglobulin classes. After the antigen binds to the B-cell receptor, the secreted form of δ chain is shut off instead of being secreted in large amounts, as is the case with IgM. The cells enter the pathways of plasma cells (which do not have IgD except for IgD-secreting plasma cells, where the μ gene is deleted) or of memory cells in secondary lymphoid organs, cells in which IgD expression is lost. Both IgD and IgM receptors can transduce signals by the same mechanism; however, signals mediated by engagement of surface IgD may differ from those mediated by surface IgM—e.g., only the latter cause death of cells by apoptosis.

After exposure to aggregated oligomeric or antigen-cross-linked monomeric secreted IgD, 10 to 15% of human peripheral blood T cells (Tδ cells) exhibit receptors for IgD and form rosettes with

IgD-coated erythrocytes. These receptors are upregulated by interleukin-2 (IL-2), IL-4, and gamma interferon on the Tδ cells. Both human $CD4^+$ and $CD8^+$ T cells can express IgD receptors, which can ligate membrane IgD during antigen presentation by B cells to facilitate the responses of antigen-specific T and B cells.

With use of sensitive assays, IgD has been found in all of the mammalian and avian species tested and is conserved across species, which suggests an evolutionary advantage. The role of IgD (especially the secreted form) is not fully understood; it has even been thought sometime that IgD might have no function.

Properties of IgE

Immunoglobulin E (IgE) is defined by the presence of the epsilon heavy chain in the structure. IgE exists as a monomer and is the least abundant antibody isotype in plasma, present at levels (about 100 ng/mL), approximately 300-fold lower than that of IgG in a circulation and accounts for 0.002% of total immunoglobulin. IgE has a very short half-life which is less than 1 day. The reason is because some proportion of the circulating amount is continually removed and destroyed in endosomes.

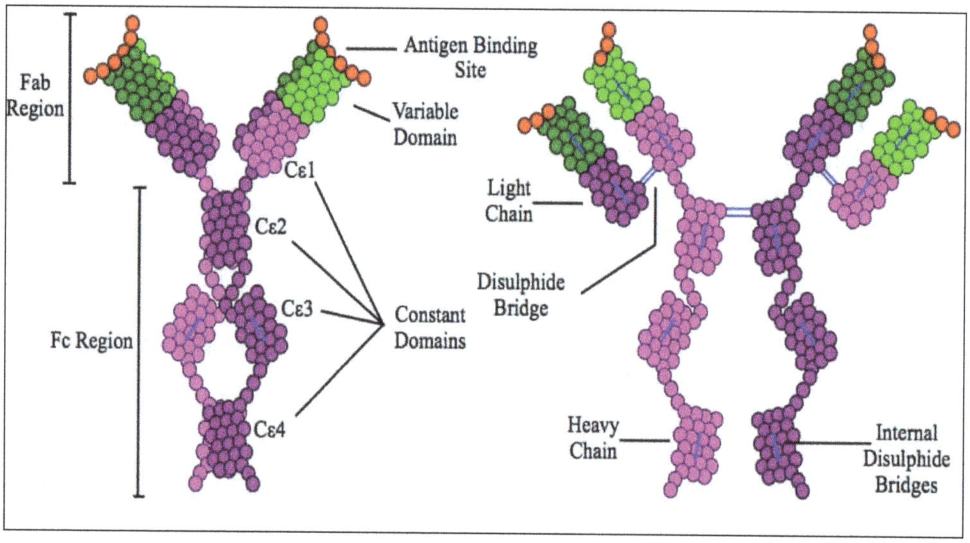

Despite the low concentrations in the circulation, IgE is extremely biologically active. This is because although the concentration of IgE in the circulation is very low, IgE antibodies bind to high-affinity receptors present on the surface of mast cells and basophils, so that these cells may be highly sensitive to allergens.

FceRI is a high-affinity receptor specific for IgE present at a high density on tissue-resident mast cells and basophils. Because of this high-affinity interaction, almost all IgE produced by B cells is bound to mast cells or basophils, which also explains the low concentration found in circulation.

Cross-linking of the FceRI-bound IgE leads to cellular activation, resulting in immediate release of preformed granular components such as histamine and tryptase and subsequent production of lipid mediators (prostaglandins and leukotrienes) and cytokines (IL-4 and IL-5).

In addition, the expression of the high-affinity receptors is upregulated during allergen-induced rhinitis in humans, probably by IgE itself. This concludes that the concentration of circulating IgE does not reflect its true activity.

Under some conditions, IgE reaches pathological levels. Clinical diseases associated with unusually high elevation of serum IgE concentrations (>1,000 IU/mL) include atopic diseases (asthma, allergic rhinitis, atopic dermatitis, urticaria), parasitic diseases, cutaneous diseases, neoplastic diseases, and immune deficiencies.

Chapter 3

T-Cell and B-Cell

A T-Cell is a type of white blood cell produced in the bone marrow and maturing in the Thymus. A B-Cell is also a type of white blood cell which matures in the bone marrow. Their difference lies in their reaction to the introduction of an antigen. T-Cells trigger the B-Cells to develop plasma cells and activate T killer cells that are ultimately responsible for killing and removing the pathogen. This chapter has been carefully written to provide a better perspective of T and B lymphocyte cells.

T-Cell

T-cells are needed for effective B-cell response. Most of the antigens evoke response from both kinds of T-cells but one kind of response may dominate. Immune system is teamwork with a well thought out plan and clear roles assigned to each member. Immune responses have broadly three phases which seamlessly segue into each other. Each phase is succeeded by another one wherein the specific responses have a greater role to play.

In the late phase > 96 hrs, there is practically little innate immunity component and memory lymphocytes form. The formation of effector cells and memory cells for T-cells is in many ways similar. While class switching make B-lymphocytes produces different classes of Ig, in case of naïve T-lymphocytes, the place, type and severity of infection dictates their differentiation into various types of T-cells.

Table: Effect of neonatal bursectomy and thymectomy on the development of immunological competence in the chicken.

All Xirradiated after birth	Peripheral blood lymphocyte count	Ig conc.	Antibody	Delayed skin reaction to tuberculin	Graft rejection
Intact	14800	++	+++	++	++
Thymectomized	9000	++	+	-	+
Bursectomized	13200	-	-	+	+

The neonatal bursectomy and thymectomy performed separately on the chickens was a very valuable approach which not only confirmed the existence of two kinds of lymphocytes, it also pointed out the respective roles of B and T lymphocyte in the phenomenon like antibody production, hypersensitivity and graft rejection. It also pointed out that some of these phenomena depends upon both B-cells and T-cells. With hindsight, we can also see that no antibodies appear in blood without Bursa, removal of thymus also diminished their concentration. So, T cells are involved in antibody production somehow.

The existence of Bursa in chicken thus was very valuable in the design of many similar studies. After almost a decade's search, it was realised that in mammals, the B-cell production was in bone marrow itself. Mammals thus, would not have been a good system as even T-cells are produced in bone marrow before migrating to thymus for undergoing thymic education.

In fact, as we know, bone marrow is the source of pluripotent stem cells or haematopoetic stem cells which differentiate into all cells present in the blood. T-cells clones are also selected from a pre-existing population upon exposure to antigens. Thus, most of the features of clonal selection theory.

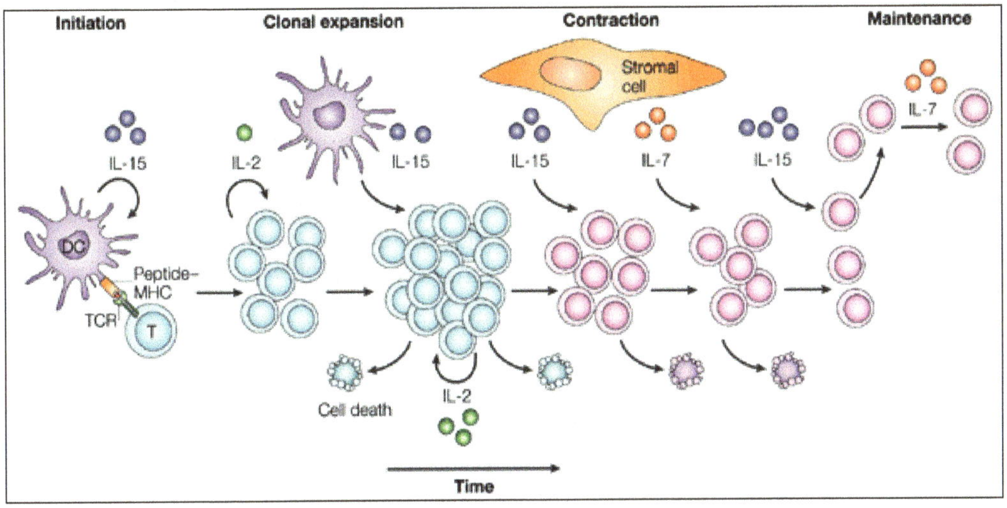

Clonal expansion and development of memory T-cells.

Selected by the antigens (as presented by an APC). Clonal expansion of T-cells takes place. Two kinds of effector T-cells are produced: T_h (helper T-cells) and T_c (cytotoxic T-cells). Just like in case of B-cells, some memories T-cells are also produced. The first version of memory T-cells retain homing and chemokine receptors of naïve cells and recirculate to the lymphoid tissues.

This clonal reservoir allows swift T-cell response. While most of the memory cells are rested, a small % keep dividing all the time. Perhaps constitutively produced cytokines may be responsible. IL-15 has been implicated in maintaining CD8 memory T-cells. The extent of cell division is such that the number of any antigen specific T-cells remain more or less constant.

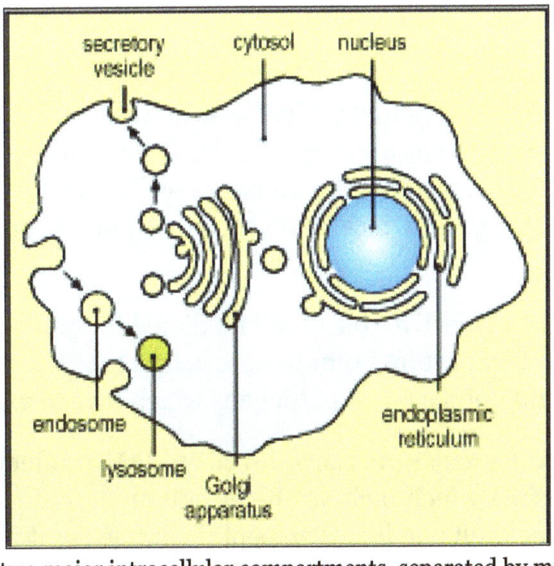

There are two major intracellular compartments, separated by membranes.

In fact a memory cell corresponding to Th cells appears faster and precedes appearance of memory B-cells. This is necessary as B-cells need T_h cells for full activation. Memory T-cells have been more difficult to isolate and characterize. All the assays for effector T-cells take several days which can allow memory T-cells to revert to effector Tcells.

	Cytosolic pathogens	Intravesicular pathogens	Extracellular pathogens and toxins
Degraded in	Cytosol	Endocytic vesicles (low pH)	Endocytic vesicles (low pH)
Peptides bind to	MHC class I	MHC class II	MHC class II
Presented to	CD8 T cells	CD4 T cells	CD4 T cells
Effect on presenting cell	Cell death	Activation to kill intravesicular bacteria and parasites	Activation of B cells to secrete Ig to eliminate extracellular bacteria/toxins

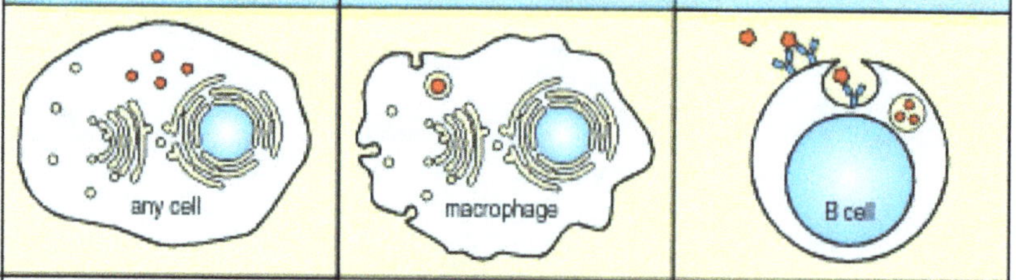

Pathogens and their products can be found in either the cytosolic or the vesicular compartment of cells.

Cell mediated immunity primarily may have evolved to mount effective response against pathogens which are intracellular. In fact, within cells pathogens can replicat into two different compartments. Viruses and some bacteria proliferate in the cytoplasm or nucleus. On the other hand some other pathogenic bacteria and parasites proliferate in endosomes or lysosomes.

Pathogens replicating in cytosol are destroyed by Tc cells. As these have CD8 as coreceptor, these are called CD8 T-cells. These are Tc cells. The other class of pathogens require response by another class of T-cells. Presence of CD4 coreceptor molecule results in their being called CD4 T cells. These are T_h cells.

As the name suggests, T_h cells play a helping role. These activate other cells and/or cooperate with other cells of the immune system. On functional basis, T_h cells are further divided into T_h1 cells and T_h2 cells. T_h1 cells are also known as inflammatory T-cells. These activate macrophages which in turn kill intravesicular pathogens. T_h2 cells, on the other hand cooperate with B-cells to result in humoral response.

It may be useful to look at the respective role of cell mediated immunity and humoral immunity in combating typical pathogens. Important point to note is that both T_h2 or T_h1 cells are required for humoral immunity to fully function against pathogens which invade extracellular fluid in tissues.

T_h2 cells in fact are required to transform naïve B-cells to IgM producing B-cells. These cells are also involved in class switching which includes B-cell production of neutralizing and opsonising subtypes of IgG. Cytosolic location results in the peptide antigens (of pathogens) being presented to CD8T cells along with MHC class I molecules. This presentation converts naïve CD8 T-cells into Tc effector cells.

Peptides from pathogens located in macrophages and dendritic vesicles provoke differentiation of T-cells into Th1 cells. In turn Th1 cells activate macrophages and also provoke B-cells to produce opsonising ab so that opsonised extracellular pathogens undergo phagocytosis. Thus T-cells participate in primary response itself and also generate memory T-cells as a result of encounter with the antigen. Memory cells differ from naïve cells but require exposure to quickly become effector T-cells.

Superantigens and T-Cell Response

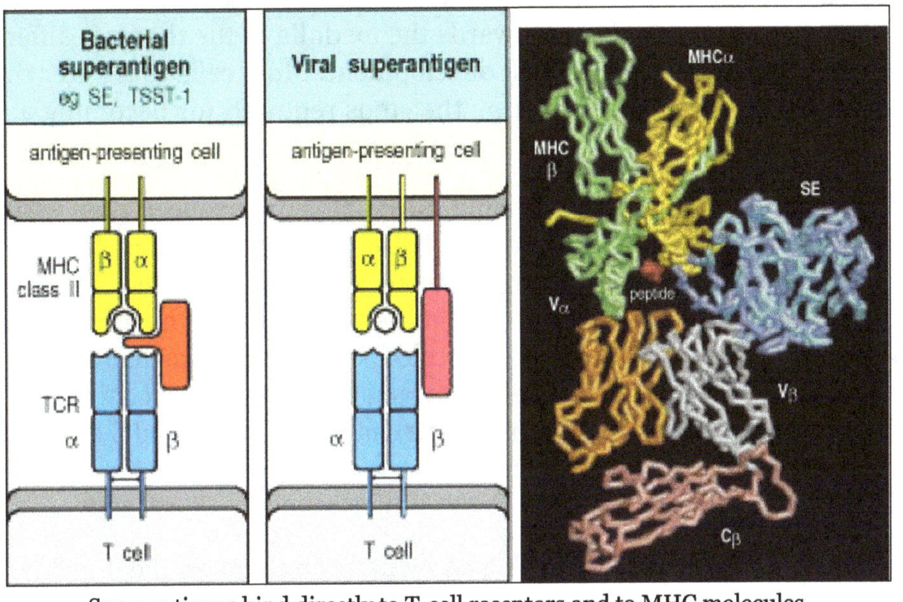

Superantigens bind directly to T-cell receptors and to MHC molecules.

Superantigens are produced by many bacteria, mycoplasms and viruses. These manage to provoke immune response which are helpful to them rather than host animal. T-cells respond to these without MHC antigens involvement. These are not processed to produce peptides for T-cell response.

These as such bind to outside of a MHC class II molecule with some other antigenic peptide bound to it. These superantigens are also able to bind to Uβ region of many TcR. Each superantigen is specified for one more of Tβ gene segments.

A superantigen can bind to about 2-20% of T-cell population. The immune response however is nonspecific, releasing cytokines from responding CD4 T cells. The result of a very high cytokine concentration results in systemic toxicity and immunosuppression. Bacterial superantigens include staphylococcal endotoxins and toxic shock syndrome toxin-1.

The best characterized viral superantigens are mouse mammary tumor virus superantigens. Superantigens, however, have proved useful in design of experiments with animals which have given insight into deletion of T-cells against self-antigens in thymus and for learning how viruses use these to facilitate their transmission.

T-Cell Development

T-cell development occurs in the thymus; the thymic microenvironment directs differentiation as well as positive and negative selection. Lymphoid progenitors which have developed from hematopoietic stem cells in the bone marrow migrate to the thymus to complete their antigen-independent maturation into functional T cells. In the thymus, T cells develop their specific T cell markers, including TCR, CD3, CD4 or CD8, and CD2. T cells also undergo thymic education through positive and negative selection.

The thymus is a multi-lobed organ composed of cortical and medullary areas surrounded by a capsule. T cell precursors enter the subcapsular cortical areas, where they encounter networks of cortical epithelial cells (the thymic stroma) and undergo a period of proliferation. As they differentiate, they move from the cortex towards the medulla of the thymus; different microenvironments within the thymus direct T cell development. Most cells that enter the thymus die by apoptosis without successfully completing the steps required for becoming a mature naive T cell.

When progenitor cells begin to express CD2 but have not yet rearranged their TCR genes ($CD2^+$ $CD3^-$), they are double negative for CD4 and CD8 ($CD4^-$ $CD8^-$), the markers for Th and Tc lineages. Of the double negative cells in the thymus, about 20% have rearranged gdTCR, about 20% have very homogenous ab TCR, and 60% are committed to becoming the majority of mature αβ T cells. These cells next express the adhesion molecule CD44, then the α chain of the IL-2 receptor (CD25). $CD44^{low}$ $CD25^+$double negative T cells rearrange TCR β chain. β chain rearrangement begins with D-J joining, followed by V-DJ joining. The chances of successful β chain rearrangement are increased by the presence of two DJCb gene clusters. If rearrangement in the first cluster fails, rearrangement in the second can occur.

Productive rearrangement of β chain is followed by its expression on the T cell membrane with CD3 and surrogate α chain, pTα (analogous to ë 5 in B cells). Signaling through the preT receptor causes the cells to stop rearranging β chain, undergo a period of proliferation, and begin to express both CD4 and CD8, becoming double positive T cells. Membrane CD25 is lost at this stage. Double positive cells re-express RAG-1 and RAG-2 to rearrange their α chain genes. A chain rearrangement can occur on both chromosomes and continue until the cell undergoes selection or dies, so T cells are not allelically excluded for a chain. However, even cells with two different TCR have only one which can bind self MHC with enough affinity to pass positive selection (one functional receptor specificity). Double positive ab T cells move into the cortico-medullary junction, where they undergo positive and negative selection and mature into Th and Tc cells.

T cell development is greatest during fetal development and before puberty. After puberty the thymus shrinks and T cell production declines; in adult humans, removal of the thymus does not compromise T cell function. Children born without a thymus because of an inability to form a proper third pharyngeal pouch during embryogenesis (DiGeorge Syndrome) were found to be deficient in T cells. Of several different T cell deficiencies that have been identified in mice, two complementary defects are found in SCID and nude mice. Nudemice, also called athymic, have a defective thymic epithelium and lack T cells. They are called nude because the defect also affects skin epithelium and results in lack of body hair as well as T cells. SCID mice (Severe Combined Immune Deficiency) have a thymus but cannot produce lymphocytes because of defects in enzymes (such as RAG and TdT) require for somatic recombination. Lymphoid progenitors from nude mice can develop normally when transferred into SCID mice with a normal thymus microenvironment.

Positive Selection of T-Cells

Double positive ab TCR^{low} cells must successfully undergo positive and negative selection before they can leave the thymus. Cells which have successfully rearranged αβ TCR will die in the thymus cortex if they do not bind self MHC within 3-4 days. Positive selection occurs when double positive T cells bind cortical epithelial cells expressing Class I or Class II MHC plus self peptides with a high enough affinity to get the survival signal. Negative selection occurs when double positive T cells bind to bone-marrow derived APC (macrophages and dendritic cells) expressing Class I or Class II MHC plus self peptides with a high enough affinity to receive an apoptosis signal. Note that selection occurs on selfpeptides in the thymus; MHC presents self peptides in the absence of pathogen.

Positive selection was demonstrated in radiation chimeras (also called bone marrow chimeras), mice whose hematopoietic cells had been destroyed by irradiation and replaced by hematopoietic cells from another mouse strain. Bone marrow was taken from F_1 a x b mice, progeny of an MHC^a parent and an MHC^b parent and having both MHC^a and MHC^b alleles on its cells. Individual T cells from a x b mice will recognize antigen presented on either MHC^a or MHC^b but not both, since they are MHC restricted. The irradiated recipient mice were either MHC^a or MHC^b. They had no hematopoietic cells but did have functional thymic stroma. The bone marrow cells from the a x b donor developed into mature white blood cells, including APC, B cells, and T cells that developed in the recipient thymus. When the recipient was infected with antigen, the T cells recognized that antigen presented only on APC of the host MHC type (MHC^a in MHC^a recipients and MHC^b in MHC^b recipients), even though both the T cells and the APC were a x b cells. (This can be demonstrated in vitro by exposing the T cells to MHC^a or MHC^b APC). Thus, the T cells had been positively selected to recognize host MHC by thymic epithelial cells.

Positive selection of T cells on thymic stroma was confirmed in a further experiment done in an $MHC^{a \times b}$ thymectomized mouse. The thymus was surgically removed when the mouse was very young, and the mouse was given a transplanted MHC^a thymus and MHC^{axb} bone marrow cells. Where the only cells in the mouse that were MHC^a were thymic stromal cells, the a x b cells still became restricted to MHC^a and could only respond to foreign antigen presented on MHC^a APC. Thus, thymus epithelial stromal cells determine "self" MHC for the developing T cells.

Chimeras made by injecting bone marrow MHC^a cells into MHC^b animals produced T cells MHC-restricted to MHC^b, but they could not respond to foreign antigen in the host mice because all the APC came from the MHC^a bone marrow cells. This experiment shows that the bone marrow donor and recipient must share at least one MHC allele for the immune system to be able to function normally. This is an important consideration for human bone marrow transplantation.

Transgenic mice for TCR restricted to a known MHC allele have demonstrated that T cells develop to maturity only if they recognize self MHC on the thymic epithelium. Remember that lymphoid progenitors with rearranged TCR transgenes genes will not rearrange their own (endogenous) TCR genes; all developing T cells will express the transgenic a and b chains of TCR. Flow cytometry with antibodies to the transgenic TCR, called clonotypicantibodies because they are specific for the TCR idiotype, and to CD4 and CD8 showed that if the transgenic T cells were not able to bind MHC, they failed to become single positive T cells and died in the thymus.

The ability of the developing T cell to make several a chain rearrangements increases its chances of undergoing positive selection. It has been shown that approximately 1/3 of T cells express more than one TCR. However, because the probability of positive selection is so low, these T cells with two TCR idiotypes should still have only a single idiotype that can recognize peptide on self MHC and not violate clonal selection.

MHC alleles and TCR genes are inherited independently. However, it appears that TCR V gene segments (with CDR1 and CDR 2 that bind MHC) are all able to bind some MHC alleles, which increases the chances of positive selection.

Positive selection also determines whether the T cell will become a helper or a cytotoxic T cell. Positive selection on Class I MHC will produce a CD8 Tc cell, while positive selection on Class II MHC will yield a CD4 Th cell. This can be confirmed in mice transgenic for TCR which are known to bind Class I or Class II MHC. Mice transgenic for TCR that is Class I-restricted produce only Tc cells; mice transgenic for Class II-restricted TCR produce only Th cells. Mice which cannot express Class II MHC generate only CD8 T cells; transgenic mice which produce only Class II on their thymic epithelial cells produce normal numbers of CD4 T cells. Likewise, mice that have a mutation in TAP and cannot present peptide on Class I MHC fail to produce any Tc cells, while mice that have no DM and cannot present peptide on Class II fail to produce any Th cells. Bare lymphocyte syndrome is a human immune deficiency characterized by lack of MHC expression and failure to produce the corresponding T cell type. Expression of co-receptor (CD4 or CD8) and its binding to MHC is also required for positive selection; mice expressing a defective CD4 that cannot bind Class II also fail to produce mature Th cells.

The genetic mechanism by which a cell becomes either a Th or a Tc is still under intense study. According to the instructive model, signals received through CD4 shut off the CD8 gene and cause the cell to differentiate into a Th, while signals received through CD8 shut off CD4 expression and induce Tc differentiation. According to the instructive model, the cell could go equally easily down either pathway and the first strong enough signal decides its fate. In the stochastic model, the cell is somehow randomly committed to becoming either a Tc or a Th before positive selection. If it gets the correct signal during positive selection, it proceeds down its predetermined pathway; if it doesn't get signaled through the correct co-receptor, it dies. Interestingly, one gene whose function may be involved is the mammalian equivalent of Notch, a gene first identified in Drosophila wing development and found to be involved in multiple developmental systems. Over-expression of Notch directs T cells into the Tc lineage, so it may normally be an inhibitor of the Th development. Notch expression has also recently been shown to be important for lymphoid progenitors to become T cells; lack of Notch expression results in mice with very few T cells.

The peptides presented by thymic epithelial cells also influence positive selection. Thymic epithelial cells express HLA-DO (H-2O in mice) that inhibits the action of HLA-DM in the MIIC vesicle. (Remember DM facilitates the removal of CLIP and the binding of processed exogenous peptide to Class II.) Thymic epithelial cells express more CLIP on their membrane Class II than other APC; they also present a range of other self peptides. Other evidence shows that the proteases in the thymic epithelial cells differ from those in APC elsewhere. The exact importance of these findings is still unclear.

Mice were made deficient in the a chain of H-2M (the mouses equivalent of DM) and all their Class II expressed CLIP. In these mice, only T cells specific for CLIP, not for other self peptides, were negatively selected in the thymus and the overall number of CD4 T cells produced was reduced. Cells transgenic for Class II-restricted TCR were not positively selected in these mice, presumably because the peptide specificity of the transgenic TCR was not for CLIP. This result indicated that the peptide does influence positive selection. T cells produced in the H-2M-deficient mice made a strong response to self peptide on APC from syngeneic (having identical MHC alleles) mice, so self-specific T cells (which would have been negatively selected in normal mice) survived in the mice presenting only CLIP. Presumably these self-specific T cells were positively selected due to their strong binding to Class II rather than to peptide, so that binding to syngeneic Class II bearing different self peptides still occurred.

Negative Selection of T-Cells

T cells that survive positive selection migrate further into the cortico-medullary junction of the thymus where they encounter macrophages and dendritic cells, bone-marrow derived APC with high expression of MHC-self peptide complexes. T cells which bind self peptide-MHC with high affinity at this stage undergo negative selection and die by apoptosis. Transgenic mice have been used to demonstrate negative as well as positive selection. Since not all self peptides are expressed in the thymus, other mechanisms for inducing peripheral tolerance must also exist.

Bone marrow chimeras demonstrate that bone marrow-derived APC are most important for negative selection. Bone marrow from MHC^{axb} mice is placed in irradiated MHC^{a} recipients, so that the T cells are positively selected on host MHCa thymic epithelial cells but also see donor MHC^{axb} macrophages and dendritic cells at the cortico-medullary junction during T cell development. Skin grafts from both MHC^{a} and MHC^{b} mice are tolerated by the chimeras, indicating that they are tolerant both to the host and the donor MHC + self peptide.

Negative selection to self antigen has been studied in mice expressing an endogenous superantigen. Superantigens bind TCR Vβ region and MHC outside the normal peptide-binding site and send strong signals to mature Th cells, inducing cytokine secretion and shock. The superantigen in these mice is a protein encoded by a mouse mammary tumor virus (MMTV) gene which has become integrated into the mouse genome and is inherited along with the mouse genes and expressed as self peptides. During T cell development, T cells with the Vb region which binds the MMTV superantigen undergo apoptosis. In mice expressing $Mls\text{-}1^{a}$ (one variant of the superantigen), T cells which have made TCRs using Vβ6, Vβ8.1, and Vβ9 segments are eliminated. These Vb segments occur with normal frequency on mature T cells of mice not expressing $Mls\text{-}1^{a}$. Double positive T cells in the $Mls\text{-}1^{a}$+ mice also express the Vβ6, Vβ8.1, and Vβ9 segments, but single positive cells in the thymic medulla do not, indicating that negative selection of these cells occurs late in T cell development.

The signals received during positive and negative selection must differ; otherwise all developing T cells would die before they leave the thymus. The differential avidity hypothesis proposes that the same peptide-MHC complex delivers both signals, but that the avidity of positive selection is lower (less signal is required to save the cells from death), while the avidity of the negative selection signal is higher (more signal is required to kill them). Experiments in which the avidity of signaling in thymus organ cultures was controlled by controlling the amount of peptide presented in TAP-deficient TCR-transgenic mice supported the differential avidity model by showing that increasing peptide presentation increased the number of T cells produced up to certain levels of peptide (positive selection), but increasing expression of the same peptide thereafter decreased the numbers of T cells produced (negative selection).

The differential signaling hypothesis proposes that qualitatively (not just quantitatively) different signals are delivered during positive and negative selection. Experiments to study this hypothesis use agonist peptides which stimulate T cells and slightly different antagonist peptides that deliver partial signals that interfere with T cell activation by agonist peptides. In this model, antagonist peptides could deliver signals leading to positive selection, but only agonist peptides could deliver strong enough signals for negative selection. This result was obtained with CD8 cells in thymus organ culture, but with CD4 cells antagonist peptides could not positively select. Differences were also observed in positive selection of CD4 and CD8 cells by directly cross-linking the TCR and co-receptors. Cross-linking TCR and CD8 (in the absence of peptide on MHC) results in the production of a partial signal and positive selection of CD8 T cells. CD4 T cells could be positively selected by cross-linking TCR with either CD4 or CD8, which produced signals similar to normal activating signals.

MHC molecules each present several but not all peptides; therefore, it might be supposed that expressing more MHC genes would improve pathogen antigen presentation. This effect is counterbalanced by the likelihood that about 5% of T cells positively selected by self peptide on one MHC molecule would be negatively selected by self peptide on another MHC molecule. The numbers of MHC genes are the optimal compromise between presenting more pathogen peptides and negatively selecting too many cells during development. Additional peptide-presenting capability is achieved by increasing the number of MHC alleles in the population. There are also limitations on the number of MHC proteins that can be expressed on the surface of a single cell. Limiting the number of MHC proteins expressed by each cell insures that each peptide-MHC complex will be presented enough times to send a strong enough signal (high enough avidity) to the T cell.

Monoclonal T cell tumors arise from different maturational stages of T cells and display characteristic membrane markers and locations. They can also be identified by their unique TCR gene rearrangements. Interestingly, double positive T cells do not seem to become cancerous.

γδ T-Cells

T cell development resembles that of B cells, with a few key differences. One difference is that while a given B cell expresses first IgM, then both IgM and IgD, and later IgG or IgA or IgE (all with the same light chain), T cells express either αβ (95% of T cells) or γδ TCR for their whole life span. The earliest T cells seen during fetal development express gd TCR. RAG-1, RAG-2, and TdT begin to rearrange γ, δ, and β gene segments nearly simultaneously and before segments are rearranged. δ gene segments are located within the a gene segment region. Immunologists believe that if γand δ are productively rearranged first, the cell will probably become a γδ T cell. If β is productively rearranged first and expressed on the membrane with surrogate a chain (pTα), the cell will usually go on to rearrange a chain gene segments and become an αβ T cell.

γδ T cells are not well understood. Some are not MHC restricted and have limited diversity in their TCR, and some can develop in a mouse which has no thymus. In mice, the first burst of γδ T cells migrate to the epidermis, where they are called dendritic epidermal T cells. γδ cells produced in the next wave settle in the reproductive tract epithelium. gdreceptors expressed on each of these groups of cells are homogenous, using the same V_γ and same d chain and without N nucleotides to induce diversity at the splice sites. It has been postulated that these γδ cells are specific for molecules produced in damaged cells, such as heat shock protein. γδ receptors on cells produced late in ontogeny and after birth are more diverse and have N nucleotides; in addition to epithelial sites, they also populate secondary lymphoid organs (but are heavily outnumbered by αβ T cells).

T-Cell Receptors

T lymphocytes have a dual specificity: they recognize polymorphic residues of self-major histocompatibility complex (MHC) molecules, which accounts for their MHC restriction; they also recognize residues of peptide antigens displayed by these MHC molecules, which is responsible for their specificity. MHC molecules and peptides form complexes on the surface of antigen presenting cells (APCs). The receptor that recognizes these peptide-MHC complexes is called the T Cell Receptor (TCR). Clones of T cells with different specificities express different TCRs.

The biochemical signals that are triggered in T cells following antigen recognition are transduced not by the TCR itself, but by invariant proteins (CD3, and zeta), which are non-covalently linked to the antigen receptor to form the TCR complex. T cells also express other membrane receptors that do not recognize antigens but participate in responses to antigens; these are collectively called 'accessory molecules'. The physiologic role of some accessory molecules is to deliver signals to the T cells that function in concert with signals from the TCR complex to fully activate the cell.

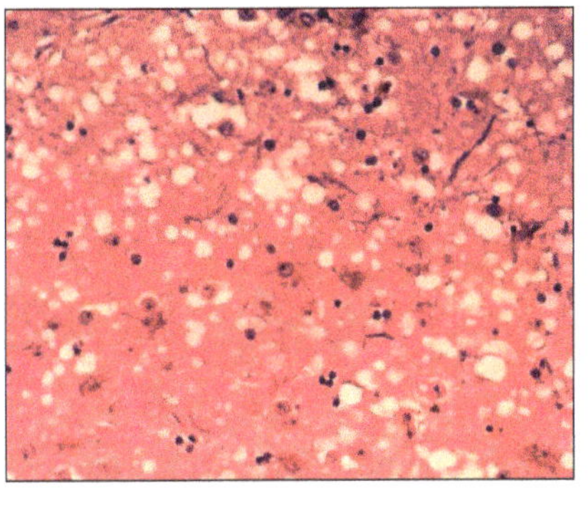

Prion-affected tissue: This micrograph of brain tissue reveals the cytoarchitectural histopathologic changes found in bovine spongiform encephalopathy. The presence of vacuoles, i.e. microscopic "holes" in the gray matter, gives the brain of BSE-affected cows a sponge-like appearance when tissue sections are examined in the lab.

The antigen receptor of MHC-restricted CD4 helper T cells and CD8 cytolytic T cell is a heterodimer consisting of two transmembrane polypeptide chains, designated alpha and beta, covalently linked to each other by disulfide bonds. Each alpha and beta chain consists of one variable domain (V), one constant domain (C), a hydrophobic transmembrane region, and a short cytoplasmic region.

The V regions of the TCR contain short stretches of amino acids where the variability between different TCRs is concentrated, and these form the hypervariable or complementarity-determining regions (CDRs). The recognition of peptide-MHC complexes is mediated by CDRs formed by both the alpha and beta chains of the TCR.

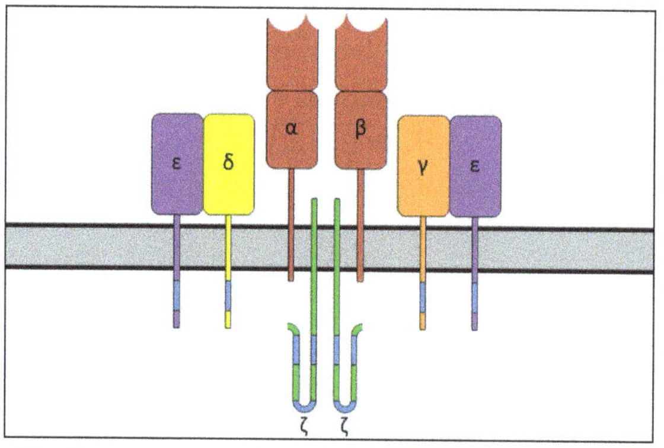

T cell receptor: T cell receptor consists of alpha and beta chains, a transmembrane domain, and a cytoplasmic region.

B-Cell

The B lymphocyte (B-cell) is one of the most important cells of the body. They form a part of the adaptive immune response by producing antibodies and presenting antigens to T cells. Once activated they will either mature into plasma cells or memory B cells.

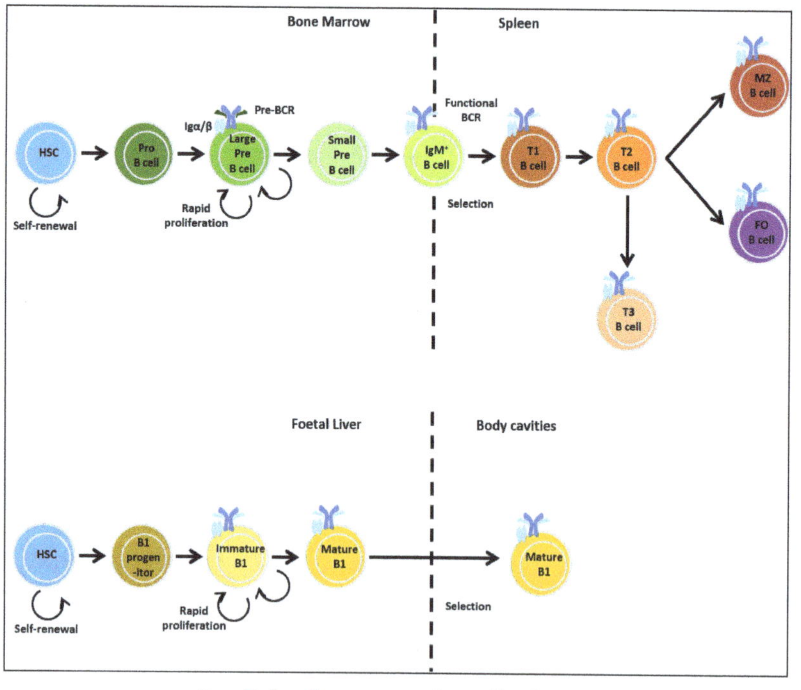

B-cell development and B-cell subsets.

Early B cell development and commitment to the B cell lineage occurs in the foetal liver prenatally, before continuing in the bone marrow throughout life. B cells are at the centre of the adaptive humoral immune system and are responsible for mediating the production of antigen-specific immunoglobulin (Ig) directed against invasive pathogens (typically known as antibodies). The function of B cells was discovered in the 1960s by Max Cooper who demonstrated that antibody production was completely abrogated in irradiated chickens after surgical removal of the Bursa of Fabricius (the primary site of B-cell development in birds) from which the notation 'B' cell was derived. Several distinct B-cell subsets have been defined that possess distinct functions in both adaptive and innate humoral immune responses.

B-Cell Development

Immunoglobulins consist of two identical heavy and light chains, which are joined by disulphide bonds. During B cell development, rearrangement of the Ig heavy chain occurs first, commencing with D-J recombination, which takes place in the common lymphoid progenitors (CLPs) and pre-pro B cells. This is followed by V-DJ recombination yielding a functional heavy chain protein (Igμ) in large pre-B cells. The recombined heavy chain then associates with the surrogate light chains and the Igα/β dimer to form the pre-B cell receptor (pre-BCR) which is expressed on the cell surface. Signalling through the pre-BCR drives intense proliferation and differentiation into the small pre-B cell stage. Quiescent small pre-B cells then undergo V-J rearrangement of the Ig light chain, allowing the production of a complete functional BCR with a unique specificity that is expressed as IgM on the surface of immature B cells. In a bid to prevent autoreactivity, immature B cells which encounter Ag capable of cross-linking their newly expressed BCRs are eliminated by a variety of mechanisms. After production in the BM, immature surface IgM+ B cells migrate to the spleen where they differentiate through distinct transitional B cell stages termed T1 and T2, before differentiating into long-lived mature follicular (FO) or marginal zone (MZ) B cells. Thus, B cells experience both antigen-dependent and -independent phases of selection, tightly regulated through signalling events. T3 B cells do not give rise to mature B cells, but instead represent a subset of anergic B cells which have been selected away from the B cell developmental pathway.

In addition to FO and MZ B cells, a third population of mature B cells exists known as B1 cells. B1 cells are located in a number of tissues including the spleen, intestine, the peritoneal cavity and pleural cavities. B1 cells have distinct haematopoietic origins in the in the fetal liver, and the initial wave of lymphopoiesis in the embryo appears to be skewed towards B1 B cell development.

B-Cell Responses to Antigen

Mature FO B cells recirculate between secondary lymphoid organs in search of antigen. Following cognate Ag encounter, B cells receiving T cell help can enter a couple of different developmental possibilities. Firstly the cells can undergo plasmacytic differentiation, form extrafollicular plasmablasts and form IgM secreting plasma cells. These cells do not have somatically mutated Ig genes and are short lived, but provide a rapid initial response to antigen. The second developmental possibility is the establishement of a germinal centre, a specialised structure within which B cells undergo rounds of proliferation accompanied by affinity maturation: an iterative process of Ig gene mutation and selection resulting in a B cell pool which can bind to Ag with the highest affinity. The cells also undergo class switch recombination. Memory B cells and plasma cells expressing somatically mutated and generally high affinity BCRs of switched isotypes exit the GC.

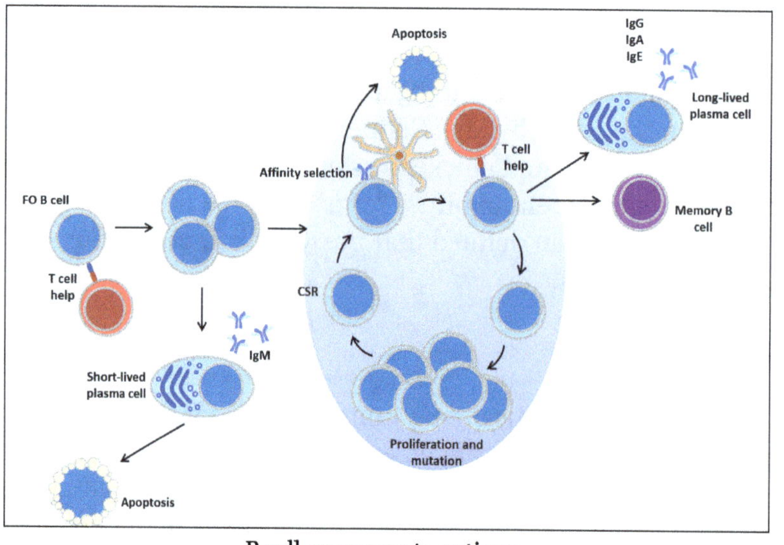

B cell responses to antigen.

Regulatory B-Cells

B cells have a positive role in priming adaptive CD4+ T cells, but not CD8+ T cells. The magnitude of CD4+ T-cell responses is reduced upon pathogen challenge in B-cell deficient or -depleted mice. B cells are also able to dampen T-cell driven immune responses, giving rise to the concept of regulatory B cells (Breg). Interleukin (IL-)10-secreting B cells with suppressive functions are referred to as B10 Bregs. B10 Bregs reduce disease severity in animal models, e.g. during experimental autoimmune encephalomyelitis (EAE), the secretion of IL-10 in mice has the effect of countering this T-cell-mediated autoimmune disease of the central nervous system. Bregs secreting IL10 or transforming growth factor β(TGFβ) have been identified in other animal models of auto-immunity, cancer and infection, supporting the concept that these cells have an important role in maintaining peripheral tolerance.

Types of B-Cell

Plasma Cell

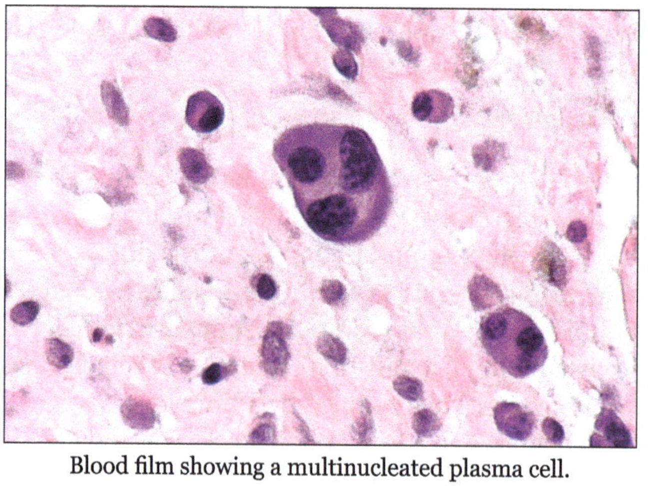

Blood film showing a multinucleated plasma cell.

Once activated, B cells can differentiate into plasma cells. Plasma cells are large lymphocytes with abundant endoplasmic reticulum, which allows them to produce large quantities of antibodies against specific antigens. They respond to signals from T cells during infection and continue to produce antibodies until the infection is controlled. Plasma cells are often found with chronic inflammation.

Memory B-Cell

Some B cells will differentiate into memory B cells when activated. These are long-lived cells which remain within the body and allow a more rapid response to future infections. If the host is re-exposed to the same antigen, these cells rapidly proliferate with assistance from T cells. This produces more B cells capable of secreting specific antibodies to the pathogen. This often means that the pathogen can be dealt with before the infection takes hold and becomes symptomatic.

T-Independent B-Cells

Most B cells require T cells to produce antibodies. However, a small number can function without T cell help. They are found within specialised sites such as the spleen and peritoneum. They are particularly important for dealing with encapsulated bacteria. Encapsulated bacteria have a polysaccharide outer layer as opposed to a protein-based one, which allows them to evade T cells. T-independent B cells can recognise these layers and produce antibodies without T cell help.

B-Cell-T-Cell Cooperation

Activation of both T, and B lymphocytes leads to two partially competing processes: cell proliferation, and differentiation, such as plasma cells, may become so specialized that they lose surface molecules such as class II and become unable to respond to regulatory signals or to proliferate. The fate of lymphocytes responding to antigen is varied. Some can persist for a long time as memory cells. The lifespan of memory cells can be over 40 years in man. Others have shorter lifespan that is nevertheless sufficient for generating effective cell-mediated and antibody responses.

Antigen-specific Activation of Lymphocytes Involves the Specific Receptors on T and B-cells

The TCR complex can transmit messages to the interior of the cell. Molecules involved are chains, and the enzyme p56lck (is a lymphocyte-specific kinase of 56kDa) that is attached to the intracellular portion of CD4 and CD8. B cells are also recognized to have a family of molecules that attached to the surface IgM and IgD, and that are o involved in signal transduction. To stimulate a T-cell hybrid an effective APC (e.g., macrophage) must carry over 60 MHC class II-antigen complexes. A weak APC, such as a class II transfected fibroblast, needs 5000 such complexes. This would suggest that, in normal circumstances, a few tens to a few hundred TCR need to be activated for effective triggering. This type of information is not available for Tdep antigens and B cells. For T_{ind} antigens, binding to a single receptor are not sufficient for triggering, but binding to tens or hundreds probably is.

T and B cells can be mitogens, a class of molecule that can activate lymphocytes in non-antigen-specific manner. The mitogens molecules such as phytohaemagglutinin (PHA) and concanavalin A (Con A) activate the T cells by binding to T cell surface molecules TCR and CD2. Superantigens are another group of antigen which activates T cells non-specifically. Most are of bacterial origin, and they include the staphylococcal enterotoxin, toxic shock syndrome toxin. These Superantigens bind to MHC class II on APCs and are recognized by TCRs, but not in the same way that MHC class II-antigen complexes are recognized. They bind only to the V beta chain of the TCR, but this sufficient to activate the T cells.

The cooperation between T and B cells is well organized. T cells are activated when they recognize antigen presented to them by an antigen presenting cell. Th cells interact with B cells that present antigen fragments to them. Activated B cells proliferate and differentiate into antibody-forming cell. Antibody is produced and various immune responses follow.

Antigen Presentation to T and B-Cells

Antigen Processing

Antigen processing refers to the degradation of antigen into peptide fragments (< 1% of antigen are involved in the immune response, the rest are degraded and excreted), which become bound to MHC I or MHC II molecules. These are the critical fragments involved in the triggering of T cells, whose receptors recognize sequences of amino acids in peptides in the MHC groove, rather than the protein shape, ('conformational determinants), which is recognized by B cell immunoglobulins. A store of antigen complexed with antibody and complement is maintained in the germinal centers of lymph nodes, adherent to follicular dendritic cells, and is used to trigger B cells.

Interaction with Antigen-presenting Cells is Essential for T-Cell Activation

The interactions between the T cells and the antigen-presenting cells (APCs) are the first interactions occurs after the antigen challenge (B cells is one type of the several kinds of the APCs), and its outcome largely dictates the subsequent course of events: if a sufficient number of $CD4^+$ T-helper (TH) cells are triggered, then the activation of B cells or the development of delayed hypersensitivity almost certainly follows. If TH cells are not triggered, or are subject to a form of immunological tolerance known as 'clonal energy', then no other immunological events follow.

How B-Cells are Activated

In order for B cells to produce antibodies, they must first be "activated". There are two ways this can be accomplished. One way definitely depends on T cell help. The other is called T cell independent activation, but it is not clear exactly how independent of Tcell help this activation pathway really is. When we talk about activation here we are talking about activation of B cells that have never before encountered antigen. This kind of B cell is usually called a naïve or virgin B cell. The rules for activating virgin B cells and for re-activating experienced B cells (those that have already encountered antigen) are somewhat different. For now we'll focus on virgin B cells that are recognizing their cognate antigens for the very first time.

B-T Cell Interaction

Interaction of B-Cells and T-Cells also Involves Multiple Surface Molecules

The interaction between B and T cells is a two way process, in which B cells present antigen to T cells, and also receive signals from the T cells for division and differentiation. The central interaction is that between the MHC class II antigen complex and the TCR; it is augmented by interactions between LFA-3 and CD2, and ICAM-1 and LFA-1. Other cell-surface molecules are also involved in interactions. The B7- 1 and B7-2 surface antigens on B cells interact with CD28 which caused stabilization of mRNA for IL-2 and other cytokines in the T cells and thereby prolongs the delivery of the activation signals. CD5 on T cells binds CD72 on the B cell surface, further promoting the interaction between the cells. CD40 delivers the most potent activating signals to B cells, even more potent than signals transmitted via immunoglobulin. Activated T cells transiently express a ligand that interacts with CD40. IL-1 and IL-6 released by some B cell enhance expression of IL-2 receptor on T cells.

Different T-Cell Subsets Activate B-Cells in Different ways

$CD4^+$ T cell can be divided into different subsets depending on their cytokine profile:

- $CD4^+$ T cells that produce IL-2 and IFN gamma but not IL-4 are designated TH1 and are responsible for delayed type hypersensitivity responses. They help B cells to produce IgG2, but not much IgG1 or IgE in the mouse.
- $CD4^+$ T cells that produce IL-4 and IL-5, but not IL-2 or IFN gamma are designated TH2. They are very efficient helper cells for antibody production, especially of IgG1 and IgE.
- Many T cells, especially in human are intermediate in their cytokine profile and are known as TH0.

T Cell Cytokine Secretion and Action is Important for B-Cell Activation

During the T-B cell interaction, T cells secrete a number of cytokines that have a powerful effect on B cells. These include IL-2, a proliferation inducer for B cells as well as for T cells, IL-4 which acts early in B cell activation of proliferation, IL-5 which in mouse but not in man is a powerful cell activator, and IL-6 which is a strong signal for B cell differentiation. T cells also produce TNF alpha and TNF beta. These molecules have also been reported to be important for B cell growth.

B cells are activated by antigen on APCs such as macrophages, in the presence of IL-4 and IL-1. This causes expression of receptors for IL-2 and other cytokines. IL-2, IL-4, and IL-5 (in the mouse) drive cell division. Only one cycle of cell division is illustrated although many cycles will usually occur. Differentiation into antibody-forming cells (AFCs) is effected by IL-4, IL-5 (in mouse), IL-6, and IL- 10 and IFN gamma.

B-T Interaction may Either Activate or Inactivate (Energize)

APC-T cell interaction may yield two opposing results, activation and inactivation (clonal anergy) of T cells. In the same way, B cells frequently become anergic. This is important because affinity maturation of B cells during the immune response, due to rapid mutation in the genes encoding the antibody variable regions, could easily result in high-affinity autoantibodies. Clonal anergy in the periphery is important device for silencing these potentially damaging clones.

T-Independent Antigens do not Require T-Cell Help to Stimulate B-Cells

The immune response to most antigens depends on both T and B cells recognizing that antigen. These types of antigens are called T-dependent (T_{dep}). There are some antigens capable of stimulating B cell without T cell help, referred to as T-independent antigens (T_{ind}). T_{ind} antigens have a number of properties in common. All are large polymeric molecules with repeating antigenic determinants. Many have the ability when they are in high concentration to activate B-cell clones that are specific for other antigens, a phenomenon known as polyclonal B cell activation; however, at low concentrations they only activate B cells specific for themselves. Many T_{ind} antigens are particularly resistant to degradation.

Primary antibody responses to T_{ind} antigens in vitro are generally slightly weaker than those to T_{dep} antigens; they peak fractionally earlier and generate mainly IgM. However, the secondary responses to T_{dep} and T_{ind} antigens differ greatly. The secondary response to T_{ind} antigens is very similar to the primary response; the secondary response to T_{dep} antigens is far stronger than the primary, and has a larger IgG component.

The Secondary Response to T_{dep} Antigens is Stronger and Induces a Greater Number of IgG-producing Cells

It seems therefore that T_{ind} antigens do not induce the maturation of response seen with T_{dep} antigens that leads to class switching to IgG and increase in antibody affinity. Memory induction to T_{ind} antigens is also relatively poor.

The mechanism by which T_{ind} antigens trigger B cells without requiring T_H cells is not fully understood. It is likely that their polymeric structure enables T_{ind} antigens to cross-link B-cell receptors; this process would be facilitated by their resistance to degradation.

Many T_{ind} antigens are products of bacteria, for example, the endotoxin, dextrans, and levans found in bacterial cell walls, and the polymerized flagellin found in flagella. There are potential survival advantages for an organism whose immune response to bacteria does not depend on complex cell interactions and may therefore be more rapid. Many bacterial antigens bypass T-cell help because they are able to induce the production of cytokines IL-1 and IL-6 and TNF alpha by macrophages. The short-lived response and lack of IgG may be due to the lack of IL-2, IL-4 and IL-5, normally produced by T cells in response to T_{dep} antigen. T_{ind} antigens often activate a subset of B cells expressing CD5.

T-Cells and B-Cells Recognize Different Parts of Antigens

Antigen enter the body is processed by cells that present the antigen in a highly immunogenic form to the TH cells and B cells. The T cells recognize determinants on the antigen different to those recognized by B cells. The T cells deliver help to appropriate B cells, which are stimulated to differentiate and divide into antibody-forming cells. Thus the two processes are required to activate a B cell:

- Antigen interacting with B cell Ig receptors.
- Stimulating signal (s) from TH cells.

T-Cell-Dependent Activation

Co-stimulating Signals are Needed for Activation

Interaction at the TCR or membrane immunoglobulin alone cannot mediate a positive activation signal for T or B cells. A number of interactions, each with a potential signaling function are involved in the cellular activation. These interactions (co-stimulatory signals) do not involve antigen-specific receptors. They may involve either secreted molecules, such as cytokines, or cell-surface molecules, which increase binding affinity and are collectively known as "adhesion molecules" (not only involve in binding, but also in signaling by their cytoplasmic domain). Cytokines such as IL-1 and IL-6, made by certain APCs, are co- stimulatory signals for T cell activation. It is likely there are other co-stimulatory cytokines. For example; IL-7 is more potent than IL-1 or IL-6 in inducing T cells to express IL-2 receptor, but it is not known whether IL-7 is produced by any APCs.

T cell-dependent activation of a naïve B cell requires two signals. The first is specific: recognition of cognate antigen by BCRs on the surface of the B cell, and the clustering of these BCRs and their associated signaling molecules. This clustering sends the "receptors engaged" message to the nucleus of the cell. However, this signal alone is not enough to activate the B cell, a second signal is required. Immunologists call this the "co-stimulatory" signal, and it is usually provided by a helper T cell (Th cell). This is where the T cell-dependent part comes in. the most important co-stimulatory signal involves direct cell-cell contact between the B cell and the helper T cell. On the surfaces of activated Th cells are proteins called CD40L. When CD40L plugs into (ligates) a protein called CD40 on the surface of the B cell, a co-stimulatory signal is sent.

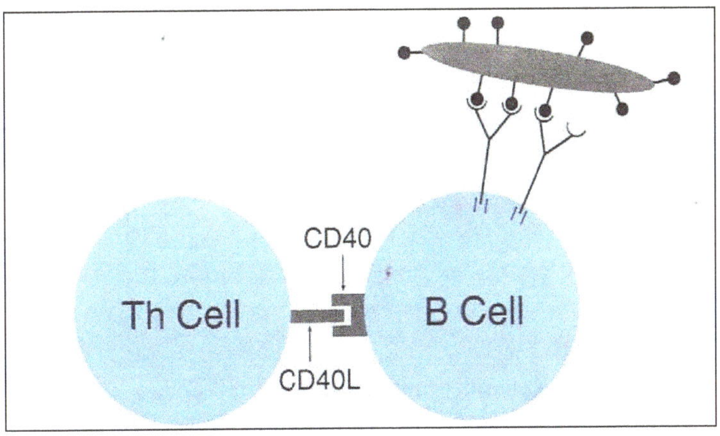

T and B-Cell Co-operation CD40 and CD40L.

The interaction between these proteins, CD40 and CD40L, is clearly very important for B cell activation; humans who have a genetic defect in either of these proteins are unable to make a T cell-dependent antibody response. Once a B cell has been activated, it remembers that experience, and when it recognizes its cognate antigen again, the requirements for re-activation are less stringent than for the initial activation. How this memory works is not clear. The latest thinking is that re- activation definitely requires recognition of cognate antigen, but that in at least some cases, physical contact between B and Th cells is not necessary.

When B cells have been activated, they express new proteins on their surfaces. One of these is the receptor for IL-2, a growth factor that stimulates B cells to proliferate. So activation of B cells makes them able to receive cytokine signals that trigger proliferation. This coupling of activation to proliferation forms the basis for clonal selection, only those B cells that have recognized their cognate antigen and have been activated will react to growth factors, proliferate, and , and form a clone of B cells with identical BCRs. The major supplier of growth factors like IL-2 is the helper T cell; another reason T cell help is usually needed for B cell activation and proliferation.

T-Cell Independent Activation of B-Cells

In response to certain antigens B cells can be activated with little or no T cell help. What these antigens have in common is that they are able to crosslink a ton of B cell receptors. In fact, clustering such a large number of BCRs appears to substitute for co-stimulation by CD40L. There are roughly two kinds of antigens that can do this. The first is an antigen that has repeated epitopes. A good example of this kind of antigen is a carbohydrate molecule of the type found on the surface of many bacterial cells. A carbohydrate molecule is made up of many repeating units, which if recognized by the BCRs together and trigger activation. Of course, this type of activation is antigen specific: only those B cells whose receptors recognize the repeated epitope will be activated. There is another quite different way that B cells can be activated, independent of Tc ell help. In this case the antigen, usually called a "mitogen", binds to molecules on the B cell surface and clusters these molecules, dragging the BCRs along.

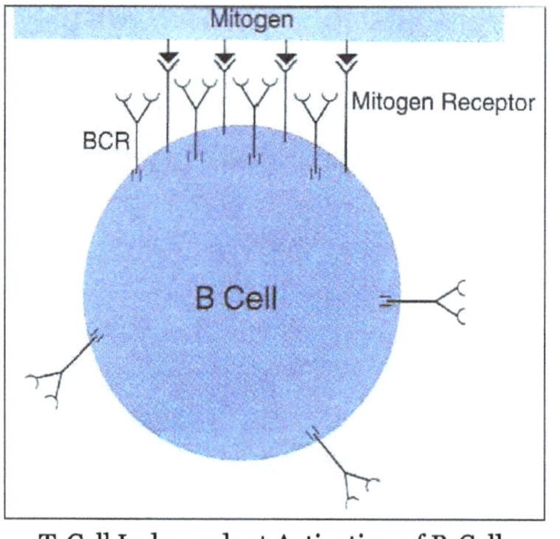

T-Cell Independent Activation of B-Cell.

This "polyclonal activation" is independent of the specificity of the BCR; the BCR just comes along for the ride. In this way, many different B cells with many different specificities can be activated by a single mitogen. Two keys are required to activate the adaptive immune system. Just as with a safe deposit box, one of these keys is specific. For B cells, this specific key is the crosslinking of B cell receptors. The second key required for activation is nonspecific in that the same key works for all B cells. In the case of T cell- dependent activation, this key takes the form of the CD40L co-stimulatory molecule that plugs into the CD40 protein on the B cell surface. By making this "two key rule", a failsafe mechanism of activation is established in which the decision to activate is made by a "committee", not by just one cell.

In careful experiments, immunologists have now shown that when a B cell's receptors are cross-linked, the B cell begins to proliferate. However, after it proliferates, the B cell won't secret any antibodies, not unless it receives a second signal. And what is this second key? For T cell-independent activation, the second key is a battle signal like the cytokine IFN-γ. What this means is that if a B cell recognizes a molecule with repeated epitopes like, for example, your own DNA, it may proliferate, but fortunately, no anti-DNA antibodies will be produced, because your immune system is not engaged in a battle either your own DNA.

On the other hand, if the innate immune system is battling a bacterial invasion and a B cell recognizes a carbohydrate antigen with repeated epitopes on the surface of a bacterial invader that B cell will produce antibodies, because battle signals generated by the innate response will function as the second signal needed for complete B cell activation. So in response to T-cell independent antigens, B cells can take their cue directly from the innate immune system and jump right into the ballet without having to wait for T cells to be activated. But there is something even more important going on here. Since T cells only recognize protein antigens, if all B cell activation required T cell help, the entire adaptive immune system would be focused only on proteins. This wouldn't be so great, since many of the most common invaders have carbohydrates or fats on their surfaces that are not found on the surfaces of human cells. Because these carbohydrates and fats are unique to invaders, they would make excellent targets for recognition by antibodies.

Activated B-Cells

The third professional APC is the activated B cell. A virgin B cell is not much good at antigen presentation, because it expresses only low levels of MHC II and little or no B7. However, once a B cell has been activated, the levels of MHC II and B7 on its surface increase dramatically. As a result, an experienced B cell is able to act as an antigen presenting cell for Th cells. The current thinking is that B cells are not used as APCs during the initial stages of in infection, because at that time they are still naïve—they haven't been activated. However, later in the course of the infection or during subsequent infections, presentation of antigen by experienced B cells is thought to play an important role. Indeed, B cells have one great advantage over the other APCs – B cells can concentrate antigen for presentation. Here is how this work:

When B cell receptor binds to its cognate antigen, the whole complex or BCR and antigen is removed from the surface and taken into the cell. Once inside the cell, the antigen is processed, bound to MHC II molecules, and transported to the cell surface for presentation.

Because BCRs have a high affinity for antigen, the B cell acts like a magnet, collecting antigen for presentation to Th cells. Because a minimum number of T cell receptor must bind to antigen for a Th cell to be activated, it is estimated that B cells have a 100-10,000 fold advantage over other APCs in presenting antigen when there is relatively little of it around. So the first time an invader is encountered, the B cells are all virgin, and the important APCs are dendritic cells and macrophages. If this same invader is encountered again, however, the experienced, memory B cells become very important APCs, because they can get the adaptive immune response cranked up quickly by concentration small amounts of antigen for presentation. Finally the cooperation between T and B cells is well organized. T cells are activated when they recognize antigen presented to them by an antigen presenting cell. Th cells interact with B cells that present antigen fragments to them. Activated B cells proliferate and differentiate into antibody-forming cell. Antibody is produced and various immune responses follow.

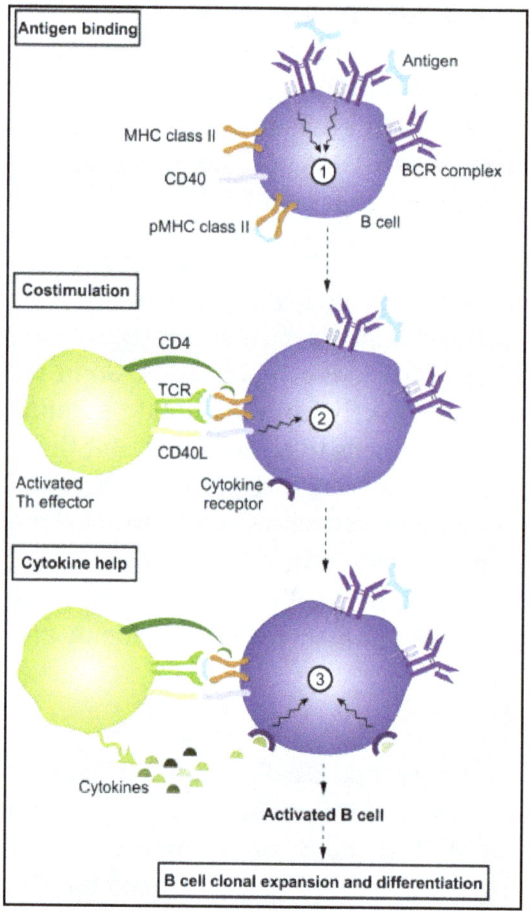

Activated B Cell.

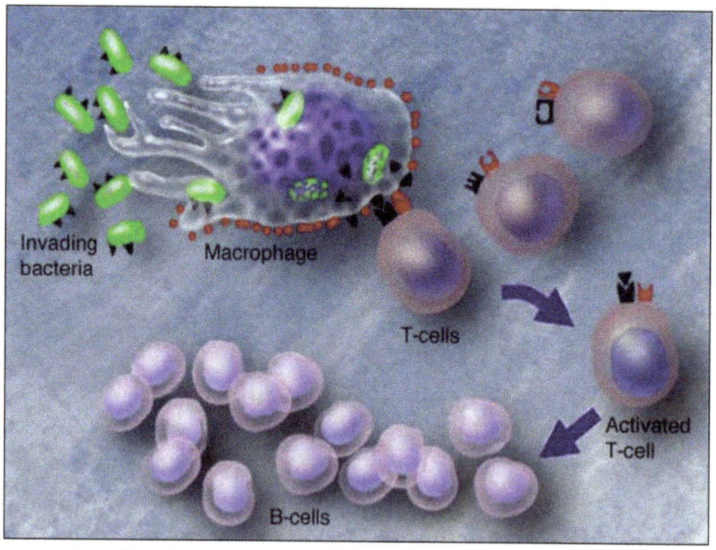

Invading bacteria are phagocytosed by the macrophage.

Figure shows the macrophage then removes the specifics identifying markers, or antigens. Of the invading bacteria. It places these antigens on its own surface. Helper T-cell lymphocytes which are specially designed to fight this invading bacteria have on their own surface a complementary antigen marker, much like a puzzle piece. When these specialty T-cell lymphocytes notice a macrophage presenting the complement antigen they initiate a response which results in B-cell proliferation.

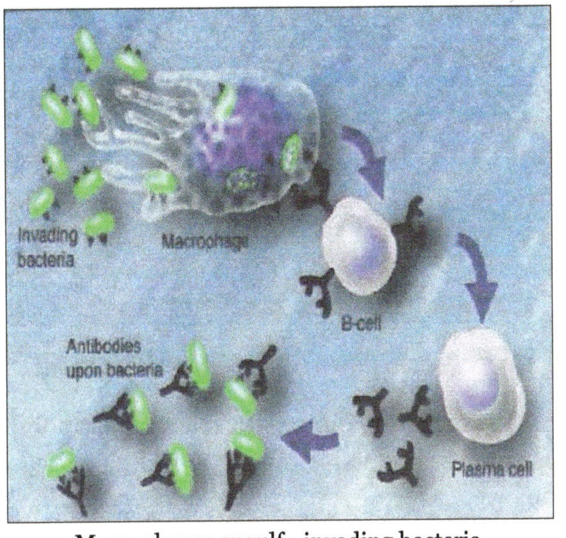

Macrophages engulf - invading bacteria.

Figure shows the macrophage will then remove the identifying markers, or antigens, from the invading bacteria and present them on its own surface. B-cell lymphocytes will notice that the antigens have been presented on the macrophage and will begin to secrete large quantities of antibodies. These antibodies bind to the invading bacteria, which will then become ineffective and more easily destroyed by macrophages.

Chapter 4
Myeloid and Complement System

Myeloid cells are blood cells that are produced in the bone marrow. They are activated upon the introduction of an antigen and play a major role in innate immunity. The complement system is a complex system of soluble proteins, enzymes and receptors that act in function together to help eliminate infectious microorganisms. Both the myeloid and complement system play a vital role in the immune system and are comprehensively covered in this chapter.

Myeloid System

All the blood cells arise from this common stem cell. Hence pluripotent haematopoietic stem cell gives rise to two different types of stem cells:

- Common lymphoid progenitor which gives rise to lymphocytes.
- Common myeloid progenitor.

Lymphoid Tissues

Lymphoid organs contain lymphocytes in the framework of non-lymphoid cells. The interaction of lymphocytes with the non-lymphoid cells is responsible for:

- Lymphocyte development.
- Initiation of the adaptive immune response.
- Maintenance of the lymphocytes.

Lymphoid organs are of two kinds:

- Central or primary lymphoid organs.
- Peripheral or secondary lymphoid organs.

Lymphocytes are generated in the primary lymphoid organs. Adaptive immune response is initiated in the secondary lymphoid organs. It is here that lymphocytes are sustained.

Primary Lymphoid Organs

There are two primary lymphoid organs: Bone Marrow and thymus. The bone marrow is an important haematopoietic tissue present in the long bones and axial skeleton. Venus sinuses are around a central artery and vein and enmeshed in developing cells. All blood cells are derived from the bone marroe stem cells. About 10% of the bone marrow cells are lymphocytes.

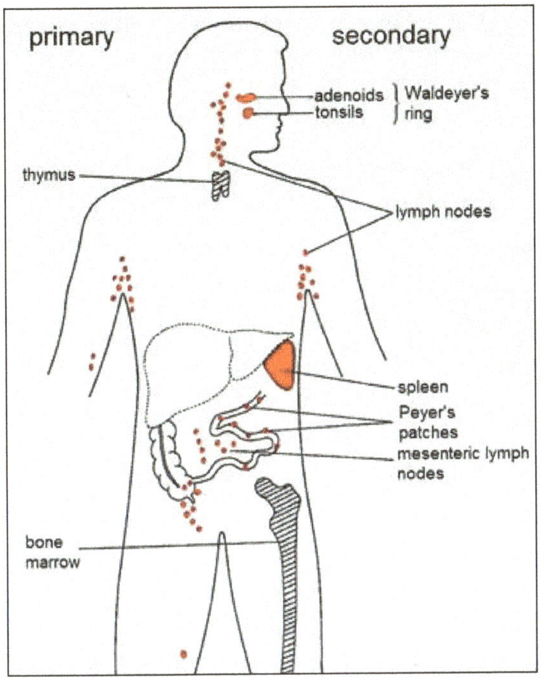

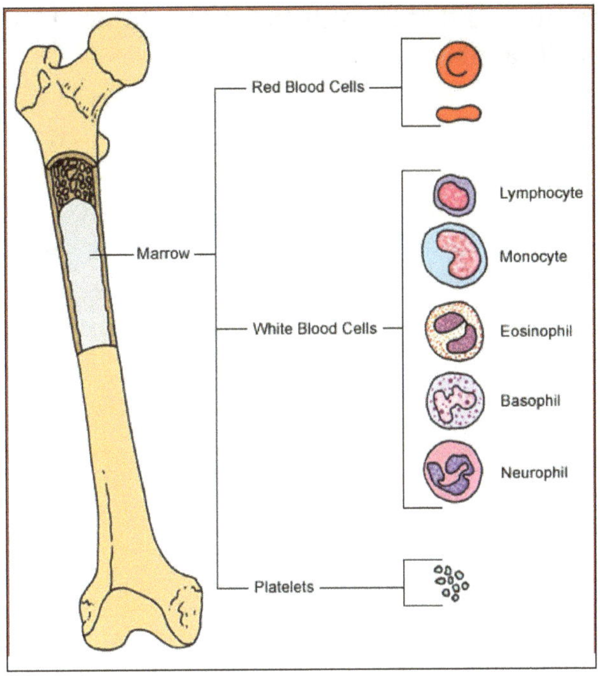

Fetal Liver and Bone Marrow: Places where B-cells Differentiate

B-lymphocyte precursors are present in the fetal liver by 8-9 week of pregnancy in humans (and 14 days in mouse). In further life, bone marrow takes over the production of B-lymphocytes. Both B- and T- lymphocytes are formed in the bone marrow. B-lymphocytes develop in the bone marrow itself. T-cells migrate to the thymus and develop there.

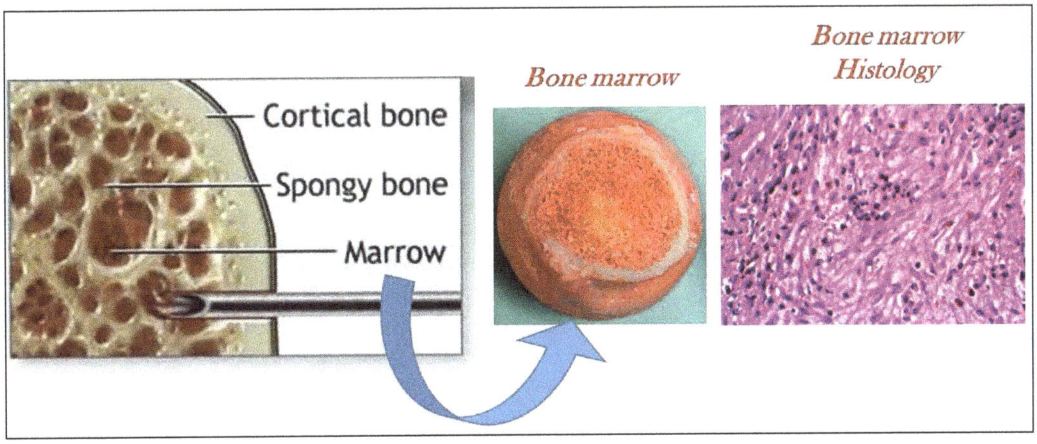

Bursa of Fabricius

Bursa of fabricius is present in birds. Despite the early belief, the effort to find a similar organ in mammals failed. Bursa of fabricius is the site where B-cells differentiate in birds. It atrophies with age. At one time, it was believed that this organ is the site of B-cell development.

Contrary to what some books claim, B-cells were so named because of this belief. It is fortuitous the Bone (marrow) also has B as its first letter. Bursa of fabricius is present at the common exit of intestinals and genitourinary tracts in birds. It has been shown that bursectomy of young chickens exhibits antibody synthesis.

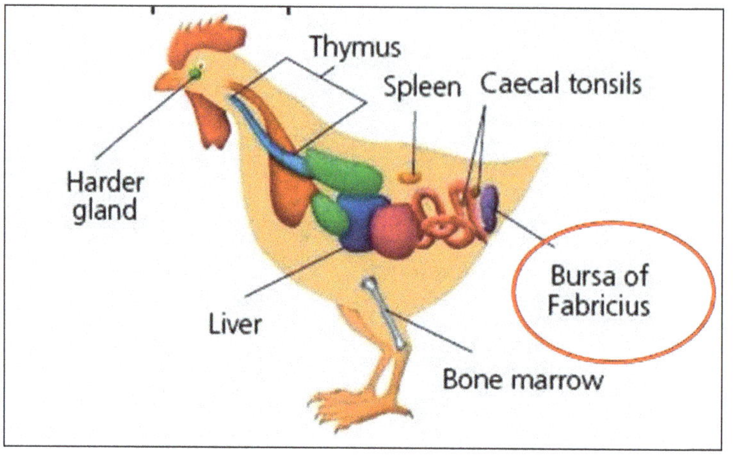

Thymus

In mammals, it is a bilobed organ and located in the thorax. Each lobe has lobules separated by connective tissue. Each lobule has inner medulla containing mature T-cells and outer cortex which is more tightly packed with immature and proliferating T-cells. Many epithelial cells present in the lobules play an important role in differentiation. In mammals, atrophy of thymus begins with puberty and decline continues with age. Pregnancy and stress both promote thymus atrophy.

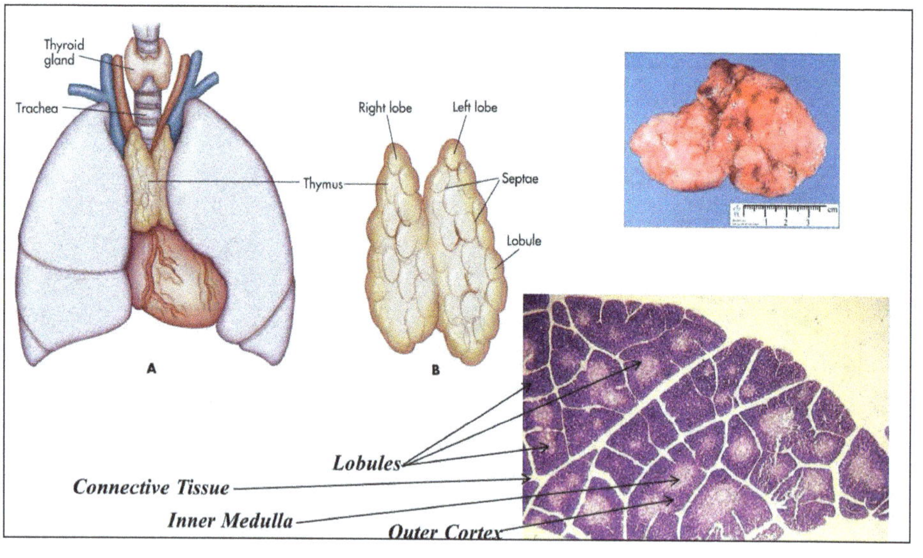

The Secondary Lymphoid Organs

Immune system actually takes place in the secondary lymphoid organs:

- Spleen,
- Lymph Nodes,

- Mucosa associated lymphoid tissues (MALT).

The mature B and T cells migrate to secondary lymphoid tissues.

Circulation of Cells

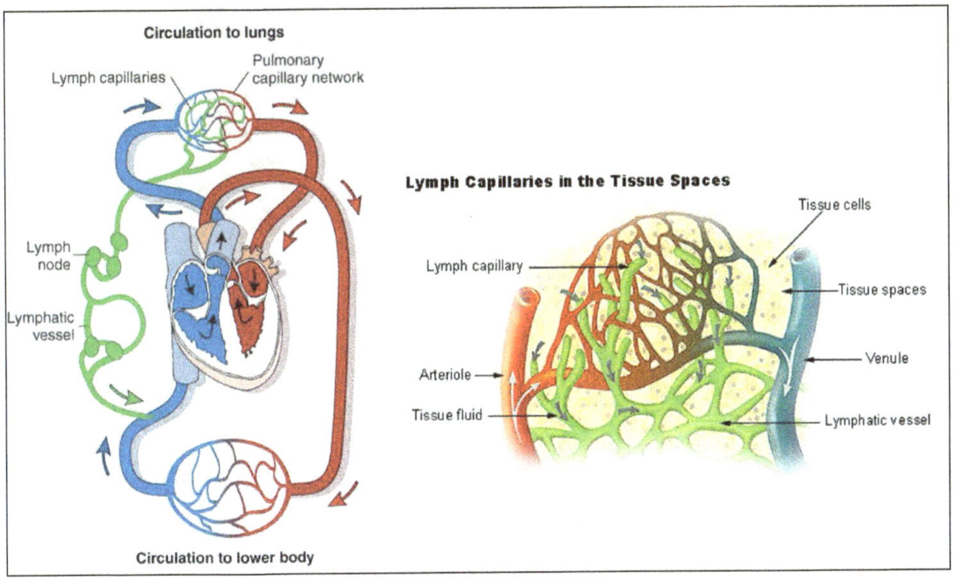

The cells (of the blood) and antibody circulate through tissues via the blood stream. The lymphatic vessels return these cells and antibody from the tissues to the blood stream. The circulation of lymphocytes between the tissue, the blood and the lymph glands then encounter with antigen or enable their participation where the immune response to the antigen has been already initiated.

The Spleen

The spleen is an encapsulated organ. Its interior, splenic pulp is spongy in appearance. The white pulp islands are situated within the red pulp which is a network of RBCs, macrophages and B-cells (Plasma cells) Spleen traps the microbes which enters it via the blood stream. It also removes the dead RBCs and immune complexes. Individuals with the spleen removed (splenectomized) are at a high risk to infections from encapsulated.

Location

To best depict the location of the spleen, we'll describe its relations. The spleen is found in the left hypochondriac region of the abdomen (left upper quadrant). More precisely, the spleen is located posterior to the stomach and anterior to the left hemidiaphragm at the level of ribs 9-10. Medial to the spleen is the left kidney; superior is the diaphragm, while inferiorly it rests directly on the left colic flexure (splenic flexure).

Although the spleen can descend as far as the pubic symphysis, as seen in mantle cell lymphoma, it normally does not move beyond the left rib arch and so is unavailable for palpation in healthy individuals.

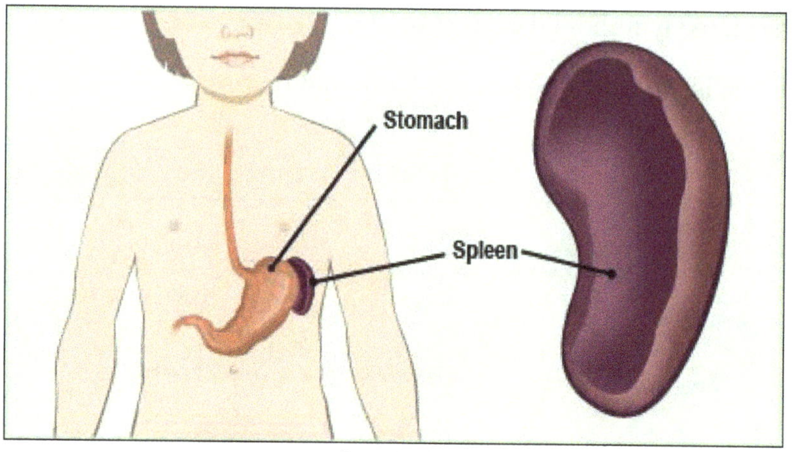

Spleen (Splen).

Anatomy

The spleen is a purple, fist-sized organ. It is wrapped by a fibroelastic capsule which allows the spleen to significantly increase its size when necessary. The spleen is an intraperitoneal organ, so all of its surfaces are covered with visceral peritoneum. Only the hilum of the spleen, the site through which the splenic artery and vein pass, is peritoneum-free.

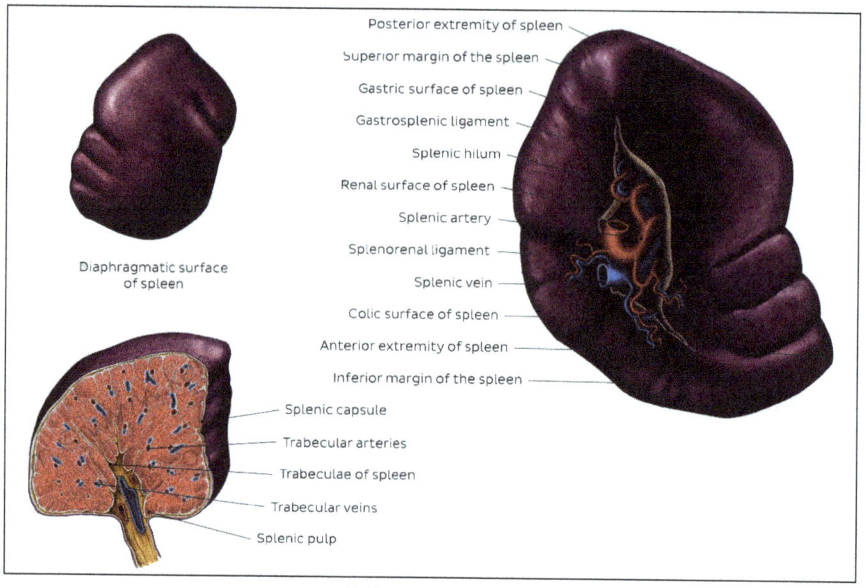

Spleen anatomy.

Organs near to the spleen leave their impressions on its surfaces which, together with spleen borders, can easily be observed and described.

- Diaphragmatic (lateral) surface leans onto the adjacent part of the diaphragm, thus it is slightly convexed to perfectly fit into the concavity of the left hemidiaphragm. This surface also shows impressions from ribs 9-11.
- Medial surface of the spleen shows three areas of impression. The colic area is the impression of the left colic flexure, the gastric area is the impression of the stomach, and the renal

area is the impression of the left kidney. The splenic hilum is found in the central part of this surface.

The spleen has three borders (superior, inferior, and anterior) as well as two extremities (anterior and posterior). The superior border bounds the gastric area, the inferior border bounds the renal area and the anterior border bounds the colic area.

Splenic Ligaments

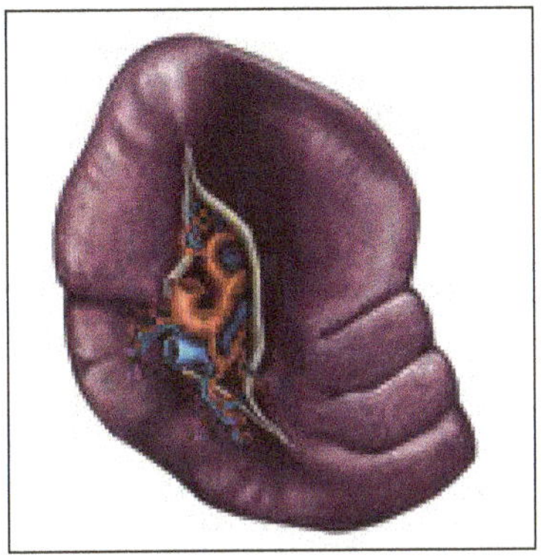

Gastrosplenic ligament (Ligamentum gastrosplenicum).

Three ligaments originating from the surrounding structures attach to the spleen. Two of these ligaments connect to the splenic hilum and are traversed by the transmitted splenic vessels. The gastrosplenic ligament connects the hilum with the greater curvature of the stomach. It contains the short gastric vessels and left gastroomental (gastroepiploic) arteries and veins. The splenorenal ligament connects the hilum of the spleen with the left kidney. It transmits the splenic artery and vein. Lastly, the spleen sits on the phrenicocolic ligament which originates from the colon and is also known as the sustentaculum lienis.

Microscopic Anatomy

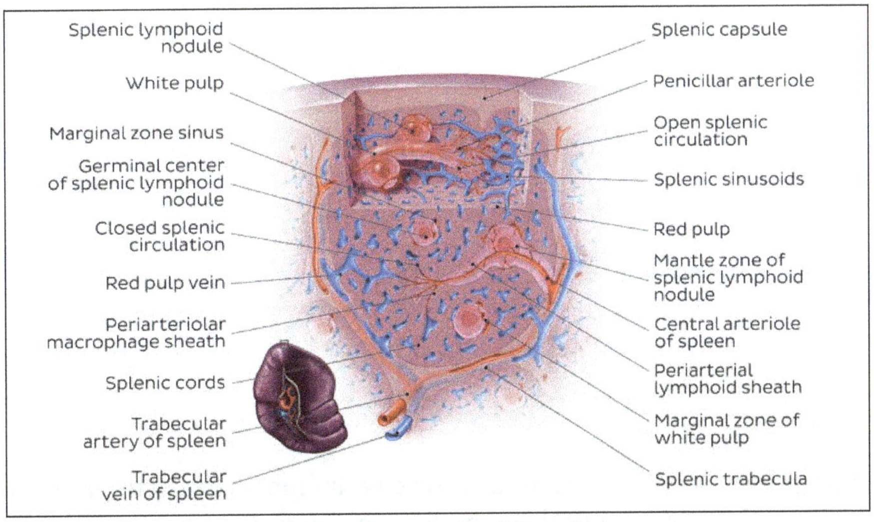

Microscopic anatomy of the spleen.

Understanding the microscopic anatomy of the spleen is important for understanding its function.

Numerous septa called trabeculae extend from the dense irregular fibroelastic connective tissue of the capsule into the parenchyma of the spleen. Both the capsule and trabeculae contain myoepithelial cells which have the ability to contract. As the spleen stores a significant amount of blood, the contraction of myoepithelial cells pumps stored blood into the circulatory system when the body is in need; for example during intense physical activity or massive hemorrhage.

The parenchyma of the spleen is called pulp. Based on the color of the pulp on fresh sections, white and red pulp can be distinguished. White pulp is the main lymphoid tissue of the spleen. It is the accumulation of lymphocytes around an arterial vessel. This aggregation of lymphocytes constitutes the lymphatic tissue known as periarterial lymphatic sheath (PALS) and it is the first to react if microbes reach the spleen through the bloodstream. The central arterial vessels in PALS nodules are branches of the splenic artery.Red pulp consists of splenic venous sinuses and cords (of Billroth), linings of splenic macrophages around the sinuses. The central artery of PALS continues from the white pulp and enters the red pulp as a capillary. These capillaries empty into the splenic cords, where macrophages phagocyte old and damaged erythrocytes. From there, blood diffuses into the splenic sinuses, thus returning to the venous circulation.

Blood Vessels

The arterial supply of the spleen comes from the tortuous splenic artery, which reaches the spleen as it travels through the splenorenal ligament. This artery emerges from the celiac trunk, which is a branch of the abdominal aorta.

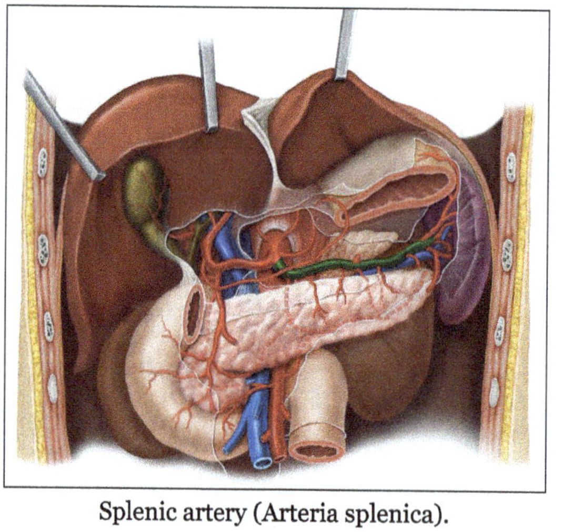

Splenic artery (Arteria splenica).

The venous drainage of the spleen occurs via the splenic vein, which also receives blood from the inferior mesenteric vein. Posterior to the neck of the pancreas, the splenic vein unites with the superior mesenteric vein to form the hepatic portal vein.

Lymphatic Drainage

The splenic lymph nodes lie at the hilum and receive lymph via perivascular and subcapsular lymphatic vessels. It is then drained to the superior pancreatic (pancreaticosplenic) lymph nodes found at the superior surface of the pancreas. From there, the lymph is drained to the celiac lymph nodes.

Innervation

The spleen is innervated by autonomic nerves from the celiac plexus, which supply the spleen with both sympathetic and parasympathetic nerves. These nerves form the splenic plexus which reaches the splenic hilum traveling along the splenic artery and its branches.

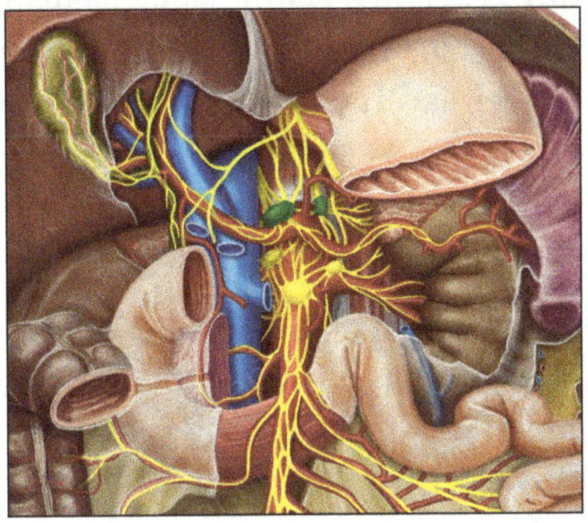

Celiac ganglia (Ganglia coeliaca).

Lymph nodes are small spherical encapsulated (2-10 nm size) structures located at various parts in the lymphatic system. Lymph nodes are found at the convergence of major blood vessels, and an adult will have approximately 800 nodes commonly cited in the neck, axilla, thorax, abdomen, and groin. These filter incoming lymph and play a role in infection as well as in malignancy.

Structure and Function

Lymph nodes are kidney-shaped and receive lymph via multiple afferent vessels, and filtered lymph then leaves via one or two efferent vessels. Nodes typically have an associated artery and vein, which terminates into a high endothelial venule (HEV). The HEV is the site of trans-endothelial migration of circulating lymphocytes due to T and B-cell endothelial surface receptors.

Lymph nodes usually range in size from 1 to 2 cm and are enclosed in an adipose tissue capsule. They are considered pathological if they lose their oval shape, if there is a loss of the hilar fat, if there is an asymmetrical thickening of the cortex and if they are persistently enlarged.

Lymph Node Structure

Capsule

The capsule of the lymph node is dense connective tissue stroma and collagenous fibers. The capsule sends trabeculae inside the lymph node, which pass inward radiating towards the center.

Subscapular Sinus

The subcapsular sinus is the space between the capsule and the cortex, which allows the transportation of the lymphatic fluid. This is also called the lymph path, the lymph sinus, or the marginal sinus. The subcapsular sinus is present beneath the capsule and is traversed by both reticular fibers and cells. It receives the afferent vessels, continues with the trabecular sinuses, and joins the medullary sinus in the medulla of the lymph node.

Cortex

The cortex of the lymph node is the layer beneath the subcapsular sinus. The cortex is formed of the outer cortex and the inner part known as the paracortex. The outer cortex layer is also named the B-cell layer, is rich in CXCR5 chemokines, and consists mainly of B-cells arranged into follicles. The immature B-cells develop into a germinal center when challenged with an antigen. Following this, resting B-cell and dendritic cells surround the germinal center to form a mantle zone. The paracortex layer, also called the T-cell layer, consists of T-cells which interact with the dendritic cells and is rich in CCR7 chemokines.

Medulla

The medulla is the innermost layer of the lymph node and contains large blood vessels, sinuses, and medullary cords. The medullary cords contain antibody-secreting plasma cells, B-cells, and macrophages. The medullary sinuses (or *sinusoids*) are vessel-like spaces that separate the medullary cords. The medullary sinuses receive lymph from the trabecular sinuses and cortical sinuses and contain reticular cells and histocytes. The medullary sinus drains the lymph into the efferent lymphatic vessels.

Function of Lymph Node

The primary function of lymph nodes is filtering interstitial fluid collected from soft tissues and eventually returning it to the vascular system. Filtering this exudative fluid allows for exposure of T-cells and B-cells to a wide range of antigens. For antigen-specific B and T cells to activate, they must first suffer exposure to antigens with the aid of antigen-presenting cells, dendritic cells, and follicular dendritic cells. These form part of both the innate immune response and play a role in adaptive immunity.

Embryology

Lymph nodes begin their development in utero as mesenchymal condensation, which later bulges to form a lymph sac. At the 13th gestational week, the T-cell region begins to develop, and by the 17th gestation week, the interdigitating reticulum cells (a subtype of T-cells) are found in the paracortical lymph node region, surrounded by lymphoid cells. B-cell regions within lymph nodes start their development at the 14th gestation week at the marginal sinus with a population of dendritic reticulum cell precursors, lymphoblasts, immunoblasts, and plasmablasts. By the 20th gestation week, incipient primary follicles are observable in the outer cortex containing lymphocytes. During the 12th and the 14th gestation weeks, lymph nodes undergo granulopoiesis and erythropoiesis to temporarily produce undifferentiated blast cells, monocytes, and macrophages.

Blood Supply and Lymphatics

Lymph Nodes of the Head and Neck

The lymph nodes in the head and neck are paired and broadly split into superficial and deep nodes.

Superficial

- Occipital nodes - At the lateral border of the trapezius muscles:
    - These drain the skin overlying the occiput.
- Mastoid (post-auricular) nodes - At the insertion of the sternocleidomastoid muscle on the mastoid process of the temporal bone:
    - These drain the posterior neck, the superior portion of the external ear and the ear canal until the tympanic membrane.
- Pre-auricular nodes - Anterior to the tragus of the ear:
    - These drain the superficial face and temporal region.
- Superficial parotid nodes - Overlying the parotid gland:
    - These drain the nose, the nasal cavity, part of the external ear canal, and the lateral orbit.
- Submental nodes - Overlying the mylohyoid muscle:
    - These drain the middle-lower lip, the floor of the mouth, and the tip of the tongue.
- Submandibular nodes - Found in the submandibular triangle, bounded by the inferior edge of the mandible and the anterior and posterior bellies of the digastric muscle, overlying the mylohyoid and hyoglossus muscles:
    - These drain the submental and facial nodes, the cheeks, the upper lip, and the marginal areas of the lower lip.
- Facial nodes - Comprised of maxillary, buccinator, and supramandibular lymph nodes:
    - These drain the mucus membranes of the inside of the cheek, the nasal mucosa, the eyelids, and the conjunctiva.
- Superficial cervical:
    - Anterior superficial cervical - Along the anterior jugular vein:
        - These drain the superficial portions of the anterior neck.
    - Posterior superficial cervical - Along the external jugular vein:
        - These also drain the superficial tissues of the neck.

Deep

- Deep parotid - Found deep to the parotid gland:
    - These drain the nasal cavity and the nasopharynx.
- Deep cervical - Found along the internal jugular vein. These are named prelaryngeal, pretracheal, retropharyngeal, infrahyoid, jugulo-digastric, jugulo-omohyoid, and supraclavicular nodes depending on their anatomical positional:
    - These drain the superficial nodes and all of the head and neck.

In terms of anatomical dissection, these more easily split into levels of the neck.

Level I

- Level Ia (submental nodes) - Anteriorly, in the midline between the anterior bellies of the paired digastric muscles.
- Level Ib (submandibular nodes) - In the submandibular triangle.

Level II

These nodes, also called the upper internal jugular nodes, are found in an anatomical area bounded by the base of the skull superiorly, the hyoid bone inferiorly, the submandibular gland anteriorly, the posterior border of the sternocleidomastoid muscle laterally and the internal carotid artery medially. The spinal accessory nerve separates the level IIa and IIb nodes.

- Level IIa (jugulo-digastric nodes) - Superficial or anterior to internal jugular vein.
- Level IIb - Deep or posterior to the internal jugular vein.

Level III

These nodes are also names the middle internal jugular nodes and are bound superiorly by the hyoid bone, the cricoid cartilage inferiorly, the anterior edge of the sternocleidomastoid muscle anteriorly, the posterior margin of the sternocleidomastoid muscle and the common carotid artery medially.

Level IV

These nodes are also named the lower internal jugular nodes and include Virchow's node. The anatomical area in which they are found is bound superiorly by the cricoid cartilage, the clavicle inferiorly, the sternocleidomastoid muscle anteriorly, and the common carotid artery medially.

Level V

These are also named the posterior triangle nodes and are bounded by the convergence of the sternocleidomastoid and trapezius muscles superiorly, the clavicle inferiorly, the sternocleidomastoid muscle anteriorly and medially and the trapezius muscle posteriorly.

- Level Va - Superior to the cricoid cartilage and include the spinal accessory nodes.
- Level Vb - Inferior to the cricoid cartilage and include the transverse cervical nodes and the supraclavicular nodes.

Level VI

This level is also named the anterior compartment and contains the anterior jugular, pre-tracheal, para-tracheal, pre-cricoid, pre-laryngeal, and thyroid nodes. It is bound superiorly by the hyoid bone, inferiorly by the suprasternal notch, by the platysma muscle anteriorly and the common carotid artery laterally.

Lymph Nodes of the Upper Limb

The deep and superficial lymphatics in the upper limb eventually drain into the axillary nodes, although there are supratrochlear and cubital lymph nodes at the level of the elbow, brachial lymph nodes, and deltopectoral lymph nodes. The drainage of the upper limbs is particular due to the presence of sentry or sentinel lymph nodes. These are usually larger than the rest of the lymph nodes and are the first to filter the incoming lymph. However, it is not uncommon for multiple smaller sentry lymph nodes to also be present.

Axillary Nodes

- Anterior nodes (pectoral) - They drain the breast and the anterior thoracic wall.
- Posterior nodes (subscapular) - They drain the posterior thoracic wall and the scapular area.
- Lateral nodes (humeral) - These are found posterior to the axillary vein and ae the primary draining nodes for the upper limb.
- Central nodes - These are found close to the 2nd part of the axillary artery and receive lymph from the anterior, posterior, and lateral nodes.
- Apical nodes - These are located near the 1st part of the axillary artery and vein and filter the lymph received from the central axillary nodes and the cephalic vein.

The apical nodes further form the subclavian lymphatic trunk, which then drains into the right lymphatic duct.

Lymph Nodes of the Lower Limb

The superficial and deep lymphatic vessels of the lower limb drain into the inguinal lymph nodes, which are in the femoral triangle. This anatomical region, also named Scarpa's triangle, is bounded by the inguinal ligament above, the medial border of the sartorius muscle laterally, and the medial border of the adductor longus muscle medially.

Inguinal Nodes

The inguinal lymph nodes split at the level (where the great saphenous vein becomes the deep femoral vein) into sub-inguinal lymph nodes below and superficial inguinal nodes above.

- Sub-inguinal nodes:
    - Superficial sub-inguinal nodes - These are found alongside the proximal saphenous vein and drain the superficial lymphatic vessels.
    - Deep sub-inguinal nodes - These nodes are commonly found alongside the medial femoral vein and collect lymph from the deep lymphatic channels of the lower limb.
- Superficial inguinal nodes - These nodes are traditionally found immediately inferior to the inguinal ligament and drain the perineal area (penis, scrotum, and perineum), the gluteal region, and part of the abdominal wall.

Iliac Nodes

- External iliac nodes:
    - Lateral external iliac lymph nodes - These are found lateral to the external iliac artery.
    - Intermediate external iliac lymph nodes - These are found medial to external iliac artery and anterior to the external iliac vein.
    - Medial external iliac lymph nodes - These are found medial to the external iliac vein.
- Common iliac nodes - They commonly arise at the level of the aortic bifurcation (4th lumbar vertebra) and extend until the level of the common iliac bifurcation (2nd sacral vertebra):
    - Lateral common iliac lymph nodes - These are found lateral to the common iliac artery.
    - Intermediate common iliac lymph nodes - These appear alongside the posteromedial common iliac artery.
    - Medial common iliac lymph nodes - These are alongside the medial common iliac artery.

Mucosa Associated Lymphoid Tissues (MALT)

The main sites of entry for microbes into the body are through mucosal surfaces. Hence, the majority (>50%) of lymphoid tissue in the human body is located within the lining of the respiratory, digestive and genitourinary tracts. Small concentrations of lymphoid tissue are also found in thyroid, breast, lung, salivary glands, eye, and skin.

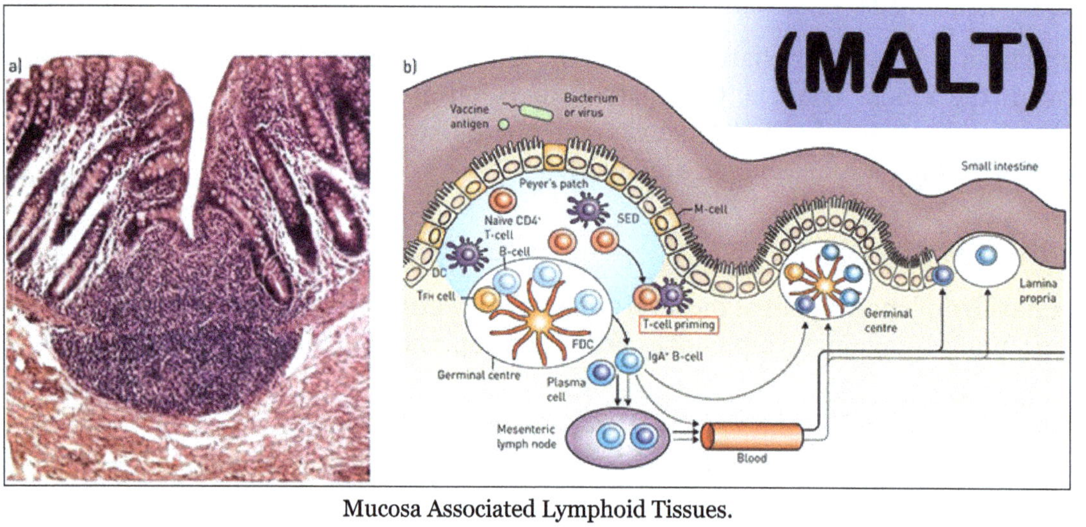

Mucosa Associated Lymphoid Tissues.

These lymphoid tissues collectively are thus referred to as mucosa associated lymphoid tissues (MALT). Examples include tonsils, the Peyer patches within the small intestine, and the vermiform appendix. Depending upon their location, they can be sub-divided into Nasal-associated lymphoid tissue (NALT), Gut-associated lymphoid tissue (GALT), Bronchus- associated Lymphoid Tissue (BALT) and lymphoid tissue associated with the genitourinary system.

Functions of MALT

The mucosa-associated lymphoid tissue initiates immune responses to specific antigens encountered along all mucosal surfaces. Diffuse lymphoid tissues along all mucosal surfaces are the sites of IgA transport across the mucosal epithelium. The primary role of GALT is to protect the body against microbes entering the body via the intestinal tract. M cells take up foreign molecules and pass them to underlying APCs, which present them in the context of class I and class II MHC molecules to T cells. The helper T cells help to activate B cells and both T and B cells can migrate to other parts of the GI tract (including salivary glands) and other MALT sites, e.g. lactating mammary glands and respiratory and genitourinary tracts, and protect these surfaces from invasion by the same microbes. The location of nasal and bronchus associated lymphoid tissues in the airway suggest that they are directly involved in handling airborne microbes. MALT tissues comprise the mucosal immune system which can function independently of the systemic immune system and are, therefore, an important aspect of immunity.

Types of MALT

NALT

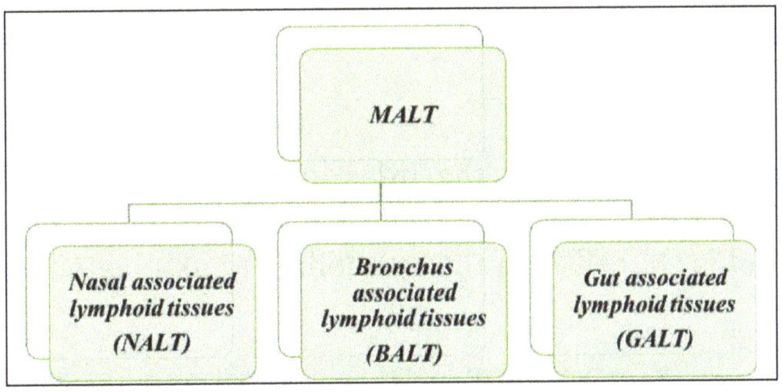

Some believe that MALT forms an interconnected system.

NALT consists of lymphoid tissue at the back of the nose (pharyngeal, tonsil and other tissue) and that associated with Waldey's ring (palatine and lingual tonsils) which is the lymphoid tissue of the neck and pharynx. NALT handles airborne microorganisms. These tissues, unlike lymphatic nodes, do not have lymphatics.

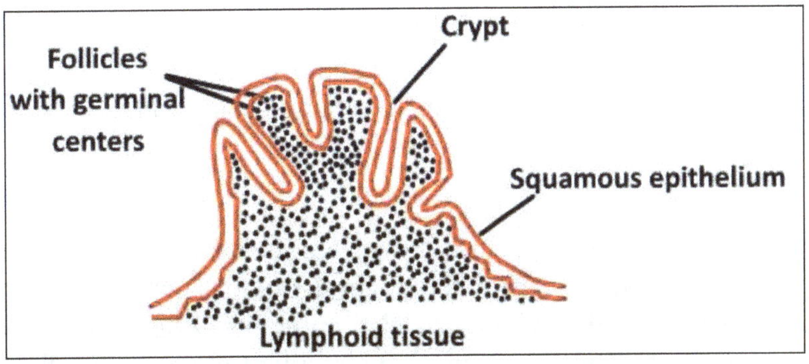

Tonsilar lymphoid tissue: Antigens trapped in the crypts are transported by M cells into the subepithelial areas where lymphocytes are stimulated via antigen presenting cells.

BALT

It consists of aggregates of lymphocytes organized into follicles present in all lobes of the lung These are situated under the epithelium mostly along the bronchi M cells transport trapped antigens to underlying cells which include lymphocytes.

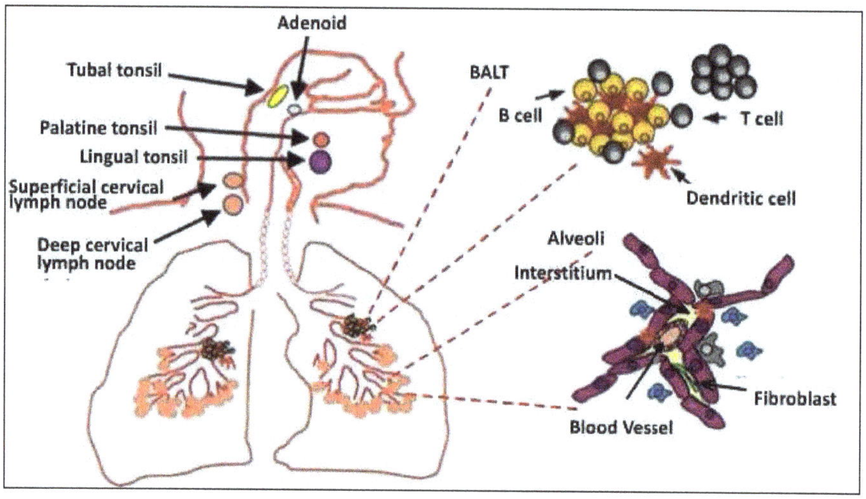

GALT

Microbes trying to enter the body via the intestinal tract are prevented by GALT. GALT has a mechanism to distinguish between harmless food and infecting microbes. This mechanism consists of microfolds (M cells) on the epithelia and associated antigen presenting cells (APCs).

A complex mechanism involving APC and T-cells decides whether the "sample antigen" results in tolerance or immune response. In fact similar mechanism also exists in BALT, Nalt and lymphoid tissues associated with genitourinary tract.

Myeloid System 2

The WBCs (leucocytes) from the common myeloid progenitor are: monocytes, immature dendritic cells, neutrophils, eosinophils, basophils. Neutrophils, eosinophils, basophils are called as granulocytes as their cytoplasmic granules have characteristic staining properties. These three are also called polymorphonuclear leucocytes as their nucleii have irregular shapes. These circulate in the blood and enter the tissue only where there is infection/inflammation.

WBCs

Immature dendritic cells enter the peripheral tissues via the bloodstream. Upon maturing there, after encountering an antigen, they migrate to the lymph nodes. Here these mature dendritic cells activate the antigen specific T-cells. Monocytes also enter the tissues via the blood stream and differentiate into macrophages. These become tissue resident phagocytic cells. Mast cells arise in the bone marrow but their complete maturation occurs in the tissues. Mast cells are important in allergic responses.

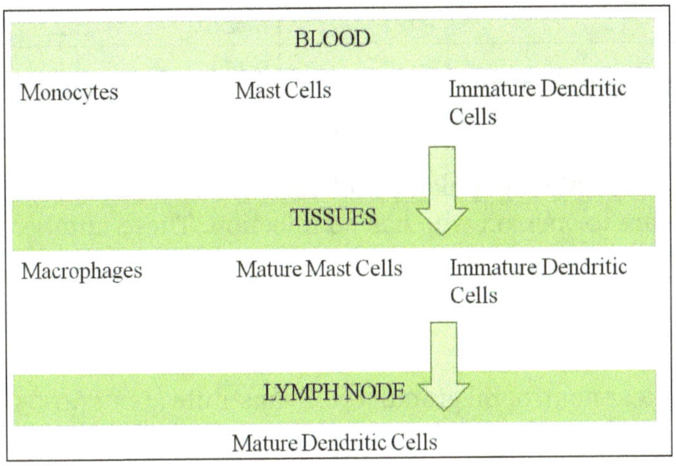

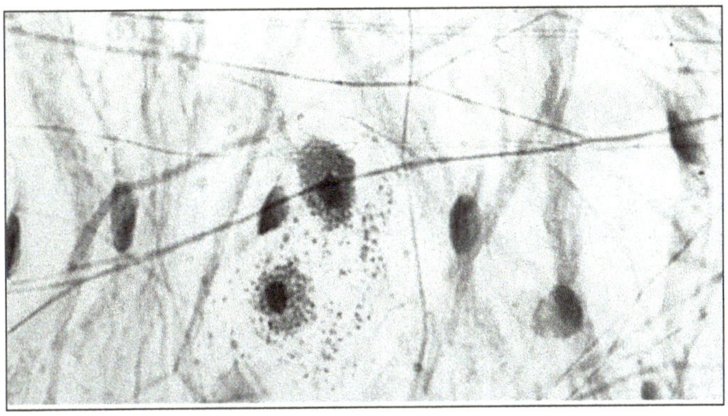

Mast Cells.

Basophils and Mast cells have both similar structure and function. Basophils circulate in the blood and their population is quite low (<0.2% of the granular leukocytes). Mast cells are found around blood vessels in the connective tissues, in gut lining, lungs and genitourinary tract. Thus, these are more or less ubiquitous in the body.

Both originate in the bone marrow. Basophils are stained with basic dyes due to their granulocytes. Both are stimulated with similar responses to release the contents of their granules. Binding of C3a and C5a (complement components), allergens to the IgE bound to the cell surfaces via Fc receptors, drugs and lectins.

Normal Leukocyte Values in Peripheral Blood

Cell Type	Cells/mm³* Median	Range	Percentage of Total Differential Count
All leukocytes (white blood cells)	7,000	4,300-10,000	100
Total neutrophils	4,000	1,800-7,200	55
Band neutrophils	500	100-2,000	10
Segmented neutrophils	3,500	1,000-6,000	45
Lymphocytes	2,500	1,500-4,000	36
Monocytes	450	200-900	6

Eosinophils	150	0-700	2
Basophils	30	0-150	1

*To calculate the number of cells/L, multiply by 10^6.

These numbers show the relative populations of various white blood cells or leucocytes. Please note that these ranges are for person who has no infection. These numbers change as a result of infection and serve diagnostic purpose.

Neutrophils

- Polymorphonuclear neutrophil granulocytes constitute about 60-75% of the blood WBC in humans/carnivores and 20-30% in ruminants.
- Their life span is of few days.

Life Span of Leucocytes

- Neutrophils → 2-5 days
- Eosinophils → 7-12days
- Basophils → 12-15days
- Monocytes → 2-5days
- lymphocytes → ½ -1 days

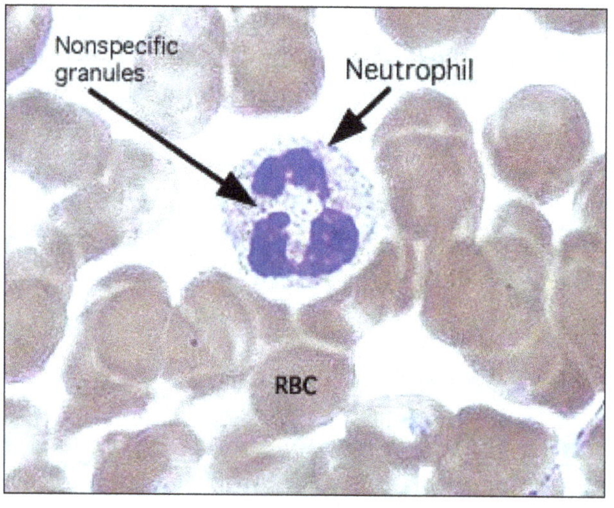

Neutrophil Maturation

Figure shows the process of neutrophil maturation begins in the bone marrow (top). After about 12 days, approximately 10% of the mature neutrophils are released into the peripheral blood, where they have a half-life of approximately 6 to 10 hours. Eventually, the neutrophils migrate into the tissues by diapedesis. The percentage of neutrophils at each stage of development (bottom) ranges from about 2% at the myoblast stage to almost 25% at the mature neutrophil stage.

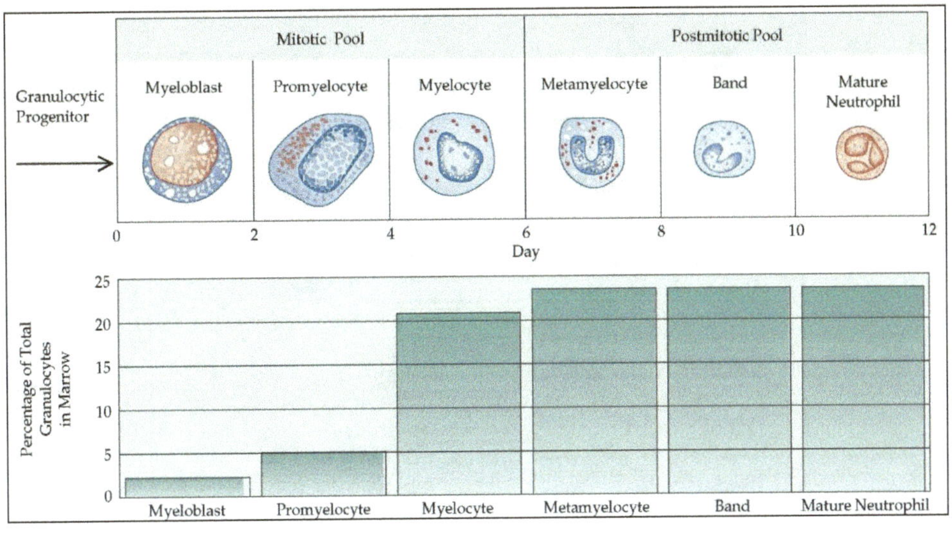

Structure of the Neutrophils

About 12 µm in diameter, their cytoplasm has two types of granules:

- Azurophils: These primary granules are electron dense and consist of enzymes like myeloperoxidase, lysozyme, elastase, beta-glucouronidase and cathapsin B.
- Specific granules: These secondary granules contain lysozyme, collagenase, lactoferrin.

Neutrophils have a small golgi and some mitochondria but no ribosomal/endoplasmic reticulum and hence cannot synthesize proteins.

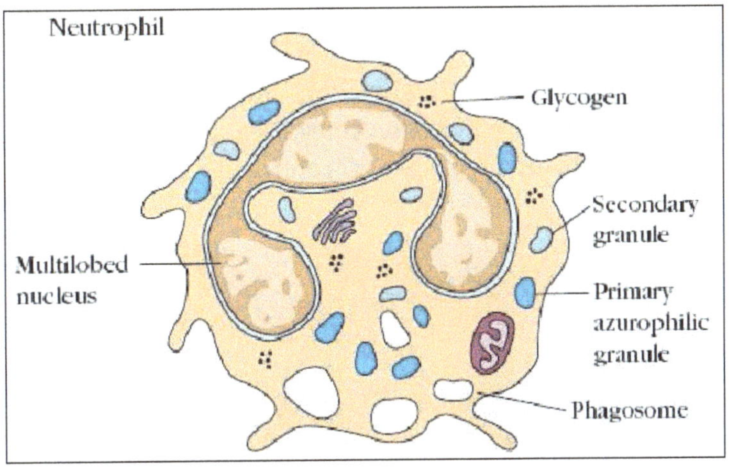

Phagocytosis by Neutrophils

Phagocytosis, destroying foreign material, is the main function of the neutrophils. The first stage of phagocytosis is chemotaxis in which neutrophils follow a chemical gradient and migrate towards the source of the chemical. Such chemicals include factors from damaged cells, leucotrienes (metabolite of an arachidonate), C5a (a peptide generated during the complement system activation) and factors from mast cells.

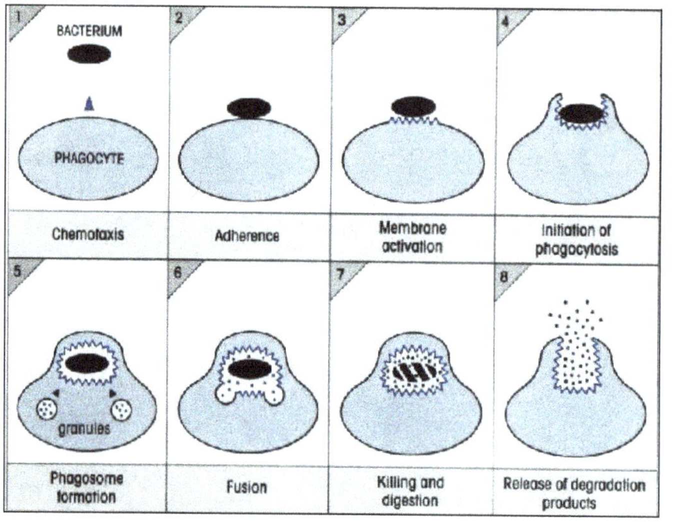

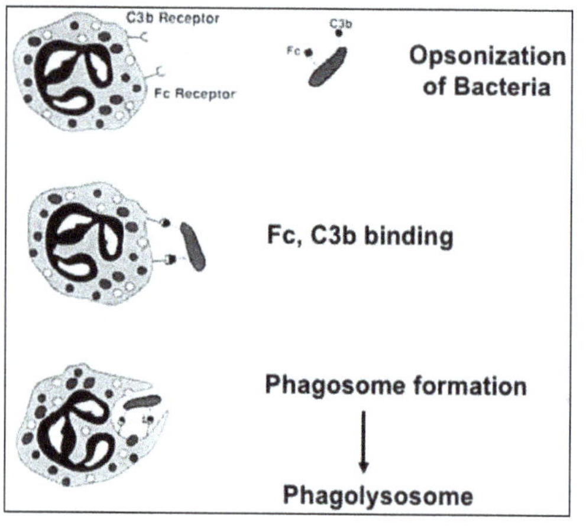

The foreign particle gets opsonized (by antibody or C3) and has lower zeta potential. This is essential as both neutrophils and foreign particles have negative charge; otherwise they will repel each other. Neutrophils have receptors for antibody or C3 and are able to bind to the opsonized particle.

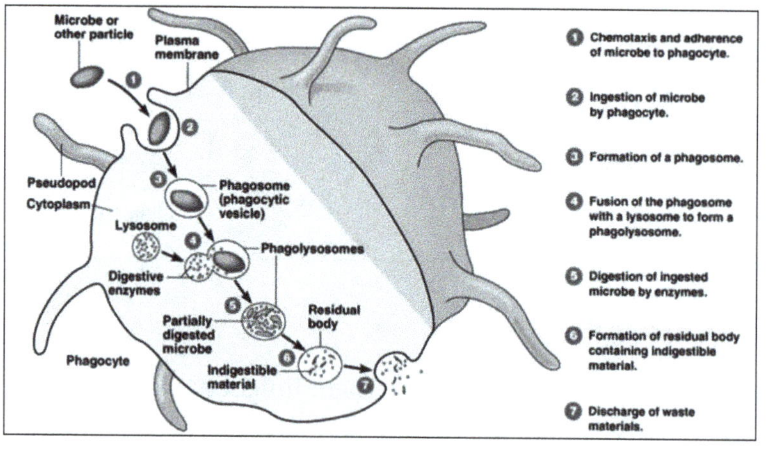

In surface phagocytosis, the particle lodged in the tissue is trapped between the tissue surface and neutrophils. This trapping facilitates contact between the particle and neutrophils. The particle gets drawn into the cell and is enclosed in a vacuole forming a phagosome. The attachment of the particle to the neutrophil membrane is a signal for the granules to move. These fuse with the phagosome to form phagolysosome.

Gram positive organisms are rapidly hydrolysed by the lysosome. Some organisms like Brucella abortus and Listeria monocytogens resist these hydrolases and may even multiply within the phagocytic cells. Infact bacteria like Streptococcus pneumoniae (which have capsule containing carbohydrates and hence is hydrophilic) has more essential requirement of opsonization before being ingested by neutrophils.

Oxidative Mechanism

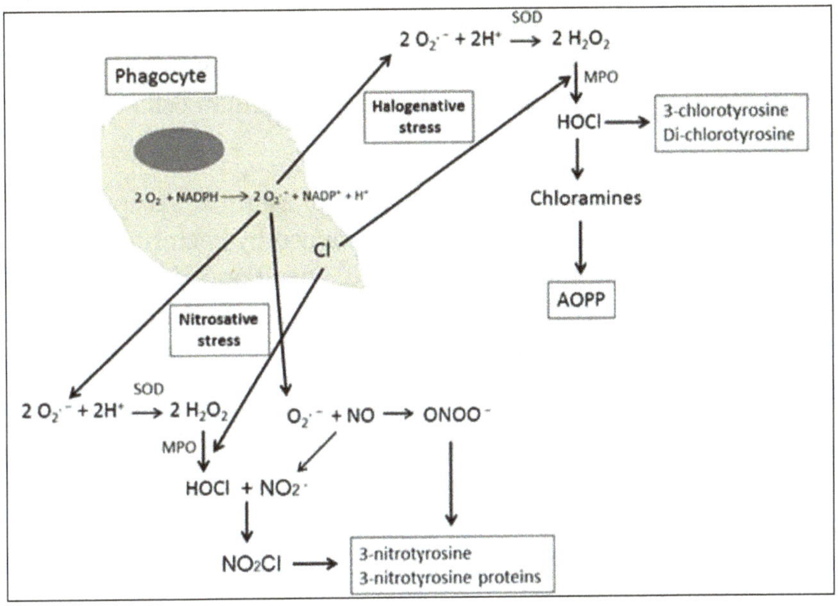

Oxidative mechanism plays an even more important role in phagocytosis by neutrophils The NADPH oxidase is the cell surface enzyme and it forms superoxide anion O_2^- . NADP$^+$ signals operation of the pentose phosphate pathway to generate energy for the cell. Superoxide dismutase (SOD) converts superoxide anion to H_2O_2.

Myeloperoxidase (MPO) converts H_2O_2 to hypochlorite ions which kill bacteria by oxidising these proteins. This reaction is similar to the one which is used in swimming pools during their chlorination. There is yet another mechanism by which neutrophils can destroy microbes. The reaction of H_2O_2 with superoxide anion generates hydroperoxide and singlet oxygen. Both can destroy bacteria by oxidation of their components.

Phagocytosis

Phagocytosis is a process wherein a cell binds to the item it wants to engulf on the cell surface and draws the item inward while engulfing around it. The process of phagocytosis often happens when the cell is trying to destroy something, like a virus or an infected cell, and is often used by immune system cells.

Phagocytosis differs from other methods of endocytosis because it is very specific and depends on the cell being able to bind to the item it wants to engulf by way of cell surface receptors. Phagocytosis won't happen unless the cell is in physical contact with the particle it wants to engulf.

The cell surface receptor used for phagocytosis depends on the type of cell that is doing the phagocytizing. These are the most common ones:

- Opsonin receptors: Opsonin receptors are used to bind bacteria or other particles that have been coated with immunoglobulin G (or "IgG") antibodies by the immune system. The immune system coats potential threats in antibodies so that other cells know it needs to be destroyed. The immune system can also use something called the "complement system", which is a group of proteins used to tag the bacteria. The complement system is another way for the immune system to destroy pathogens and threats to the host.
- Scavenger receptors: Scavenger receptors bind to molecules that are produced by bacteria. Most bacteria and other cellular species produce a matrix of proteins surrounding themselves (called an "extracellular matrix"). This matrix is a perfect way for the immune system to identify foreign species in the body, because human cells do not produce the same protein matrix.
- Toll-like receptors: Toll-like receptors, named after a similar receptor in fruit flies encoded by the Toll gene, bind to specific molecules produced by bacteria. Toll-like receptors are a key part of the innate immune system because, once bound to a bacterial pathogen, they recognize the specific bacteria and activate the immune response. There are lots of different types of Toll-like receptors produced by the body, all which bind different molecules.
- Antibodies: Some immune cells make antibodies that can bind to specific antigens. This is a process similar to how toll-like receptors recognize and identify what type of bacteria is infecting the host. Antigens are molecules that act like a pathogen "calling card", because they help the immune system know what threat it has to fight.

How does Phagocytosis Happen?

Cells have to complete some steps in order to successfully phagocytize something:

- The virus and the cell need to come into contact with each other: Sometimes the immune cell accidentally bumps into a virus in the blood stream. Other times, cells move by way of a process called "chemotaxis". Chemotaxis means the movement of an organism or cell in response to a chemical stimulus. Many immune system cells move in response to cytokines, small proteins used specifically for cell signaling. Cytokines signal cells to move to certain area in the body where the particle is found. This is common with infections that are specific to a certain area (like a skin wound infected with bacteria).
- The virus binds to the cell surface receptors on the macrophage: Remember that different cell types express different receptors. Some receptors are general, meaning that they can identify a self-produced molecule versus a potential threat (and that's about it), and others are very specific, like toll-like receptors or antibodies. The macrophage will not initiate phagocytosis without successful binding of the cell surface receptors.

  Viruses can also have surface receptors which can be specific to those on the macrophage. Viruses need to access the host cell's cytoplasm or nucleus in order to replicate and cause an infection, so they use their surface receptors to interact with immune system cells and exploit the immune response for entry into the cell. Sometimes, when a virus and a host cell interact, the host cell is able to successfully destroy the virus and stop the spread of infection. Other times, the host cell engulfs the virus, and the virus tricks the cell, gaining access to what it needs to replicate. Once this happens, the infected cell is identified and destroyed by other cells of the immune system in order to stop viral replication and infection.

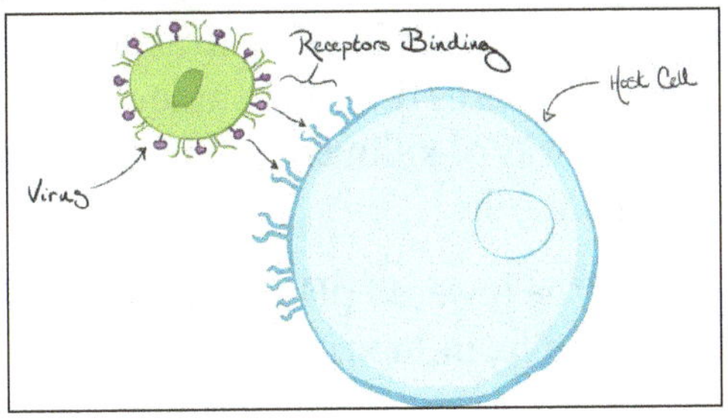

- The macrophage starts to surround the virus and engulf it into the cell. Instead of moving the large item across the plasma membrane, which might damage the membrane permanently, phagocytosis uses extensions of the cytoplasm (pseudopods) to surround the particle and enclose it in a membrane. The macrophage and virus are bound at the cell surface. The pseudopods protrude outward on either side of the virus until both sides meets and the virus is enclosed. Remember, cells are reasonably flexible and fluid.

  The macrophage and virus are bound at the cell surface. The cell pulls the virus inward, creating a pocket-like indentation without damaging the plasma membrane. Remember, cells are reasonably flexible and fluid.

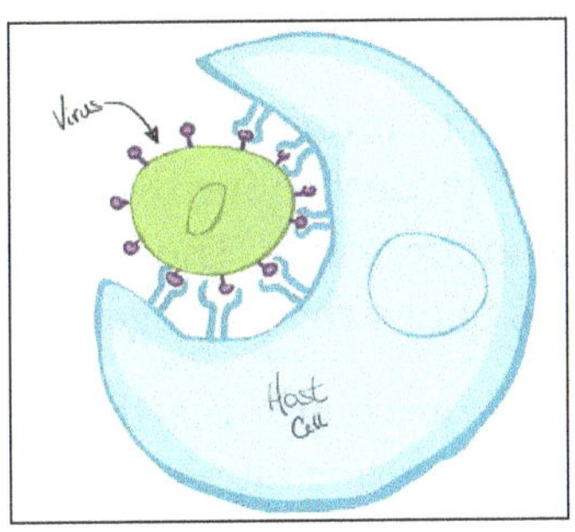

- The surrounded virus becomes completely enclosed in a bubble-like structure, called a "phagosome", within the cytoplasm: The lips of the pocket, formed as a result of the extensions of the pseudopodsIn extend towards each other in order to close the gap. This action creates a phagosome, where the plasma membrane has moved around the particle, encasing it safely inside the cell.

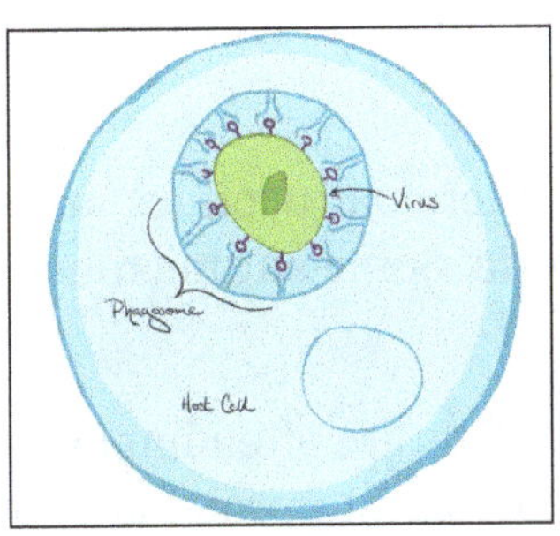

- The phagosome fuses with a lysosome, becoming a "phagolysosome": Lysosomes are also bubble-like structures, similar to phagosomes, which process wastes inside the cell. "Lysis" means "to break down", makes it easy to remember the function of a lysosome. Without fusing with a lysosome, the phagosome wouldn't be able to do anything with the contents inside.
- Phagolysosome lowers the pH to break down its contents: A lysosome or phagolysosome is able to break down the stuff inside of itself by drastically lowering the pH of its internal environment. Lowering the pH makes the environment inside the phagolysosome makes it very acidic. This is an effective way of killing or neutralizing whatever is inside the phagolysosome so it cannot infect them cell.

  Some viruses actually exploit the lowered pH to escape the phagolysosome and start replicating inside the cell. For example, influenza (the flu virus) uses the drop in pH to activate a conformational change, allowing it to escape into the cytoplasm.
- Once the contents have been neutralized, the phagolysosome forms a residual body that contains the waste products from the phagolysosome: The residual body is eventually discharged from the cell.

Phagocytosis and the Immune System

Phagocytosis is a critical part of the immune system. Several types of cells of the immune system perform phagocytosis, such as neutrophils, macrophages, dendritic cells, and B lymphocytes. The act of phagocytizing pathogenic or foreign particles allows cells of the immune system to know what they are fighting against. By knowing the enemy, the cells of the immune system can specifically target similar particles circulating in the body.

Another function of phagocytosis in the immune system is to ingest and destroy pathogens (like viruses and bacteria) and infected cells. By destroying the infected cells, the immune system limits how quickly the infection can spread and multiply. The immune system cells that perform phagocytosis can also use other mechanisms to destroy pathogens inside the phagolysosome, such as:

- Oxygen Radicals: Oxygen radicals are highly reactive molecules that react with proteins, lipids and other biological molecules. During physiological stress, the amount of oxygen radicals in a cell can increase dramatically, causing oxidative stress, which can destroy cell structures.
- Nitric Oxide: Nitric oxide is a reactive substance, similar to oxygen radicals, that reacts with superoxide to create further molecules that damage various biological molecules.
- Antimicrobial Proteins: Antimicrobial proteins are proteins that specifically damage or kill bacteria. Examples of antimicrobial proteins include proteases, which kill various bacteria by destroying essential proteins, and lysozyme, which attacks the cell walls of gram positive bacteria.
- Antimicrobial Peptides: Antimicrobial peptides are similar to antimicrobial proteins in that they attack and kill bacteria. Some antimicrobial peptides, like defensins, attack bacterial cell membranes.
- Binding Proteins: Binding proteins are often important players in the innate immune system because they competitively bind to proteins or ions that would have otherwise been

beneficial to bacteria or viral replication. Lactoferrin, a binding protein found in mucosal membranes, binds iron ions, which are necessary for growth of bacteria.

Functions of Macrophages

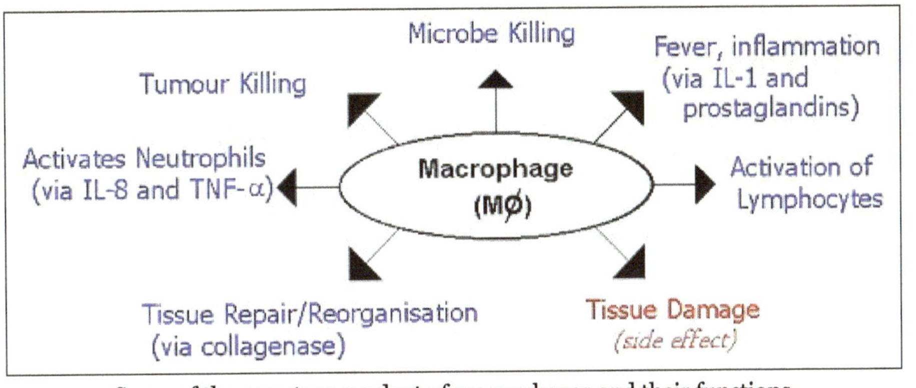

Some of the secretory product of macrophages and their functions.

- Phagocytosis: Macrophages phagocytose not only microbial products but also act on damaged cells (especially those damaged by neutrophils). Thus neutrophil action is followed by macrophage action.

  Macrophages have the following receptors. The macrophages have less intensive oxidative action and have catalase instead of myeloperoxidase. These receptors enable macrophages to act on foreign material opsonized by molecules recognized by these receptors.

  Macrophages upon encounter with microbes/damaged cells release interleukin 1. This protein causes fever, recruits neutrophils (Co-operation with neutrophils, before as well as after the macrophage encounter with foreign substances) and in general fight off invasion and repair injury.

- Inflammation: Inflammation is an extremely complex process. Macrophages release large number of substances upon reaching the site of microbial attack and are important in the healing process as well.

  Macrophages can process some antigens. Macrophages possess the capacity to synthesize proteins. This is necessary as several secreting proteins by macrophages play important roles in various biochemical processes.

Complement System

Complement system is a set of serum enzymes. It is an important illustration of how cascade system can amplify a biological signal. The major forms of complement system are: Initiating inflammation, attraction of neutrophils, opsonisation of the microbe and killing of the microbe by lysis.

The complement system is common to all vertebrates. It is one of the serum enzyme systems. It is a cascade system in which one reaction leads to form the active enzyme for the next. The result is a domino effect where a small stimulus can get amplified down the line.

The Complement Proteins

In humans, complement contains 20 soluble glycoproteins. Many of these are produced by hepatocytes and monocytes. These are constitutively present in blood. These are designed as C_1, C_2, etc. or as factor B, etc.

Why Cascade Reactions are useful?

In many cases, uncontrolled activity can have disastrous consequences. Yet when needed, it needs to be in place when required. A known example is that of blood clot formation. Blood clots can be fatal when formed inside the body. However, once bleeding starts a slow clot formation would lead to an individual bleeding to death. The strategy is to limit such activities by producing them in small quantities and with short half-lives so that high amount of activity is generated transiently.

Cascade system like complement achieves this beautifully. Complement system was discovered by Belgian scientist Bordat. It is a system whose activation results in disruption of cell membrane. There are various pathways by which complement system operates.

In the classical pathway, the trigger is an immune complex of Ag and Ab.

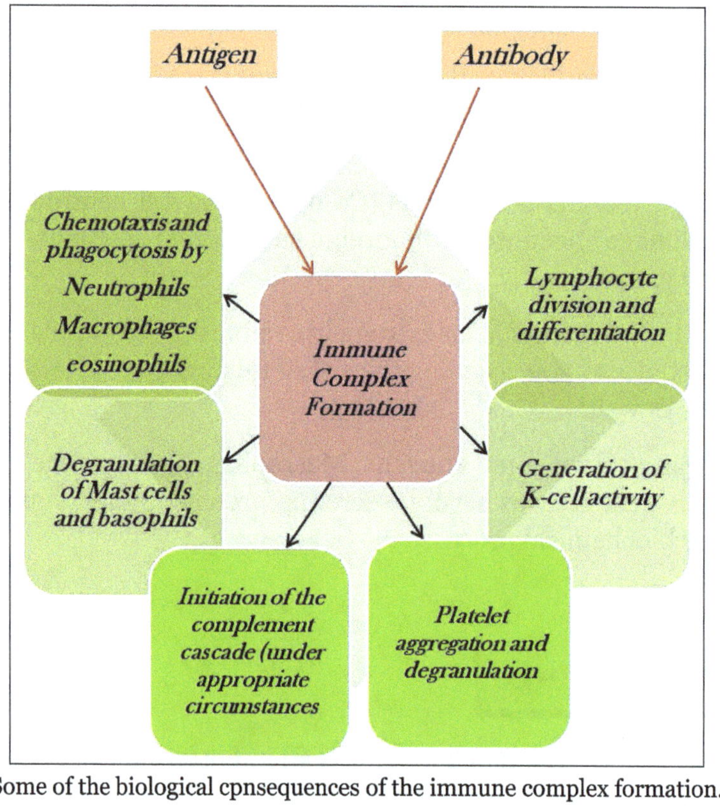

Some of the biological cpnsequences of the immune complex formation.

The properties of antibody complexed with antigen are very different from free antibody. Firstly the network of antigen and antibody creates a spatial array of molecules. Secondly antigen binding to antibody via F_{ab} generates a conformational change which leads to F_c region being more available for many interactions.

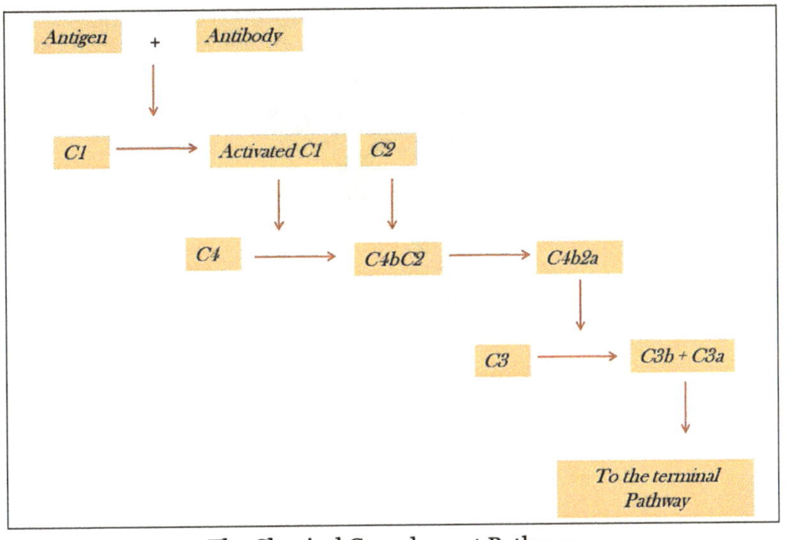

The Classical Complement Pathway.

C1 can be stimulated by IgM bound to antibody or pair of IgG bound to an antbody but located closely on the antibody surface. Although the complement components are numbered, the sequence of their participation is not in numerical order. Thus C1 acts on C4 and so on. Polymeric IgM activates complement more effectively than IgG.

Table: Components of the complement cascades.

Component	$M_r \times 10_3$	Main Sources	Concentration in serum (µg/mL)
C1q	390	Gl epithelial cells, fibroblast, macrophages	190
C1r	83	Gl epithelial cells, macrophages	
C1s	83	Gl epithelial cells, macrophages	110
C4	209	Macrophages, liver	430
C2	117	Macrophages	30
D	23.5	Macrophages, plateletes	2
B	100	Macrophages, liver	240
P	186	Macrophages	23
C3	190	Macrophages, liver	1300
C5	206	Macrophages	75
C6	95	Liver	60
C7	120		55
C8	163	Liver	80
C9	71	Macrophages	58
I	88	Macrophages platelets	34
H	160	Liver	500

This table shows components of all the pathways of the complement activation. In the classical pathway, as we have seen, only C1, C4, C2, C3 are involved.

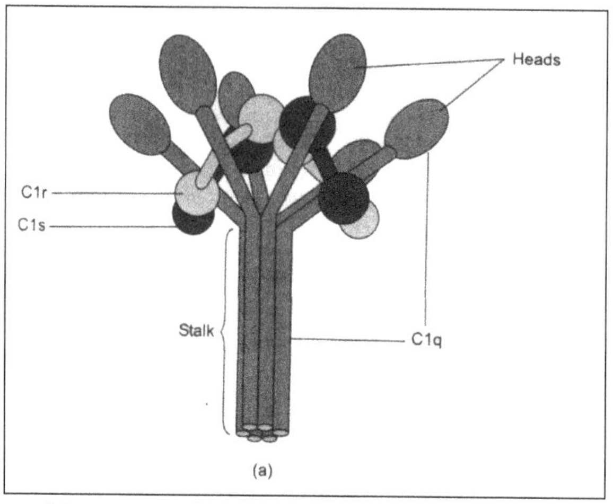

Structure of C1 macromolecular complex, consisting of C1q and $C1r_2s_2$.

The five components in the complex $C1q(C1r)_2(C1s)_2$ are bound to each other non-covalently and stabilised by Ca^{2+}. C1q itself consists of 18 polypeptide chains in which three polypeptides A, B, and C form a subunit . C1 is made up of one molecule of C1q and two molecule of C1r and C1s each.

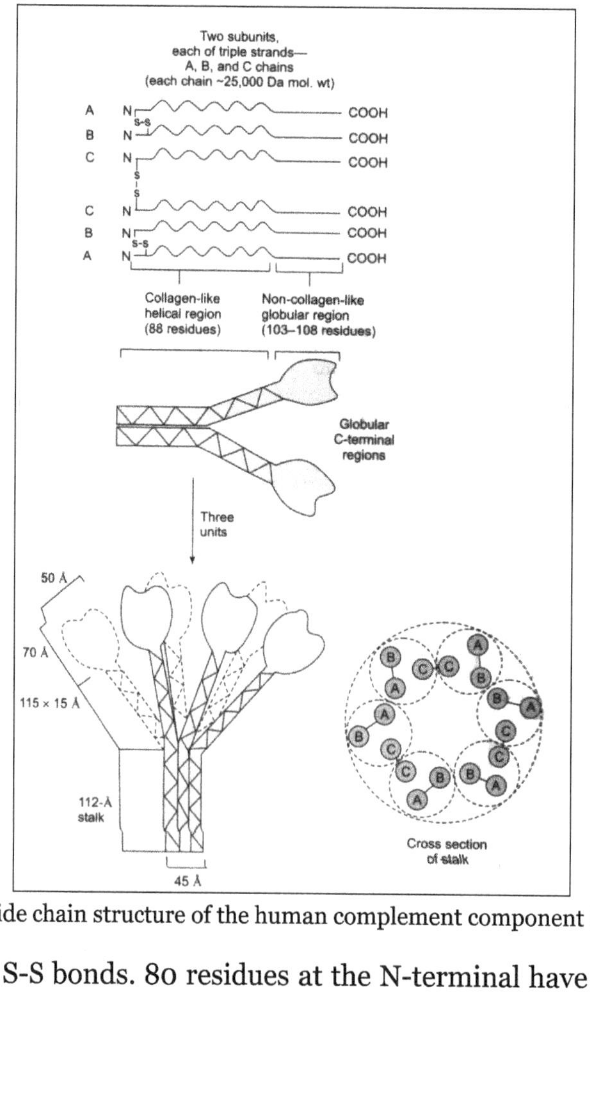

Peptide chain structure of the human complement component C1q.

The chains are linked by S-S bonds. 80 residues at the N-terminal have a collagen like structure with glycine occupying 3rd position in a sequence rich in Hydroxyproline, Hydroxylysine and of course Glycine . In fact, Collagenase can digest part of C1q .

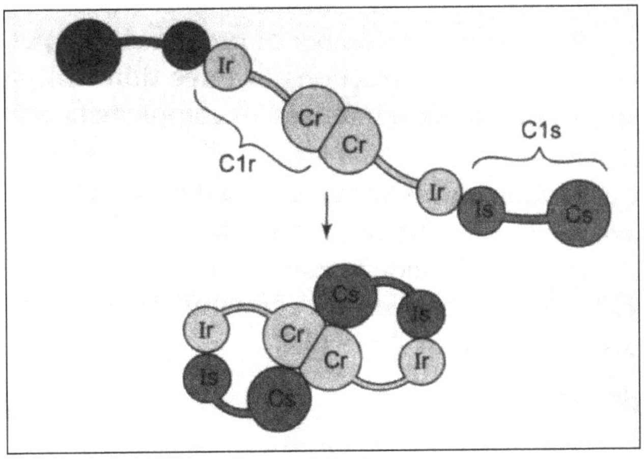

Shape of the $C1r_2s_2$.

The complex $(C1r)_2$ $(C1s)_2$ can occur as such and has an extended form. Upon binding with C1q, it turns into a more compact form twisted with shape of 8. Each C1r and C1s has one catalytic domain and one interaction domain which is responsible for binding with each other as well as with C1q. Binding of C1q brings catalytic domains of C1q and C1s together. The globular heads of C1q (the C-terminals which do not have collagen like structure!) is responsible for binding to the F_c portion of the antibody (which in turn is bound to an antigen!). Each C1q molecule has to bind to at least two F_c sites for a stable complex formation to take place. This is the molecular mechanism of the trigger which requires either 2Ig molecules closely or an (IgM bound to antigen).

Also, in circulating IgM, C1q binding sites on F_c are not exposed. Hence, unless bound to an antigen, IgM cannot initiate complement cascade reactions.

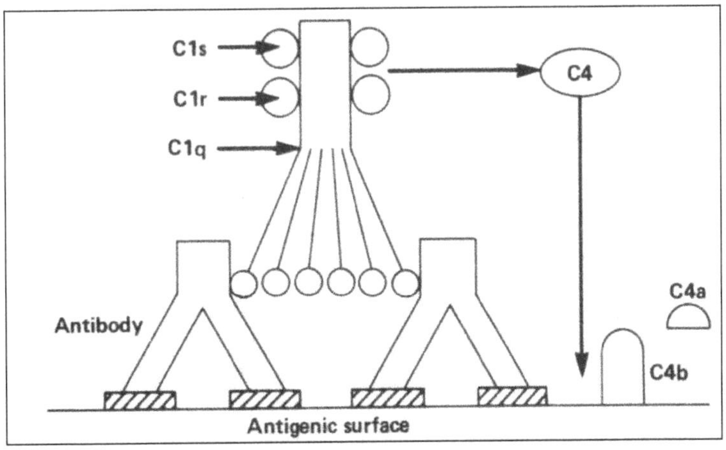

The activation of C4 by the C1 complex.

C4 is the next component which comes in play. It is glycoprotein with three chains α, β and γ. Their molecular weights are 93kDa, 78kDa and 33kDa respectively. C1r and C1s potential catalytic domains coming together results in C1r getting activated and splitting C1s. This exposes catalytic sites on C1s. C1s cleaves C4. The active C1s in fact cleaves α-chain of C4 to form two fragments: Small fragment from the N-terminal called C4a and a major fragment C4b.

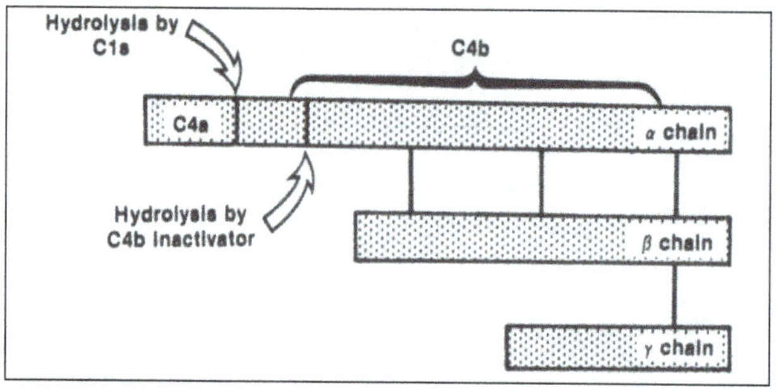

The structure of C4.

C4b is unstable in free form and can be rapidly broken down to an inactive product,

$$C4b \xrightarrow{\text{C4 bioactivator}} C4bi$$

C4b can however bind to neutrophils and platelets and lymphocytes as all these cells have the complement receptors on their surface. C4b can also continue to participate further in the complement pathway.

Activated C1s can also split C2. C2 is a glycoprotein of 117 KDa.

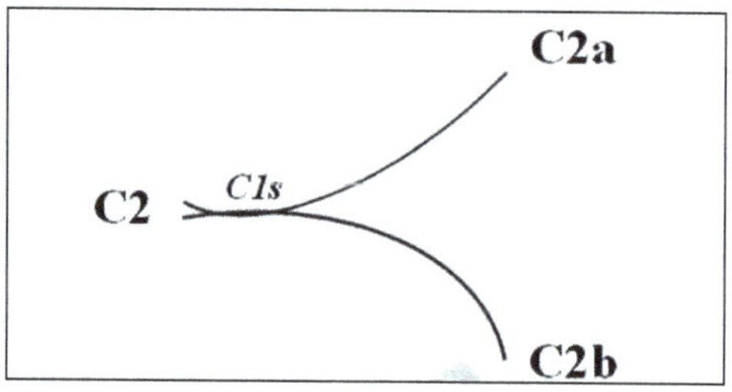

C2a can combine with C4b. C4bC2a is called classical C3 convertase as it activates C3.

$$C3 \xrightarrow{\text{C3 convertase}} C3a + C3b$$

C4bC2a can further complex with C3b to form C4bC2aC3b which is called C5 convertase.

C3

It has molecular weight of 190kDa and has one α chain (120kDa) and one β-chain (70 kDa) linked by 2 S-S bonds.

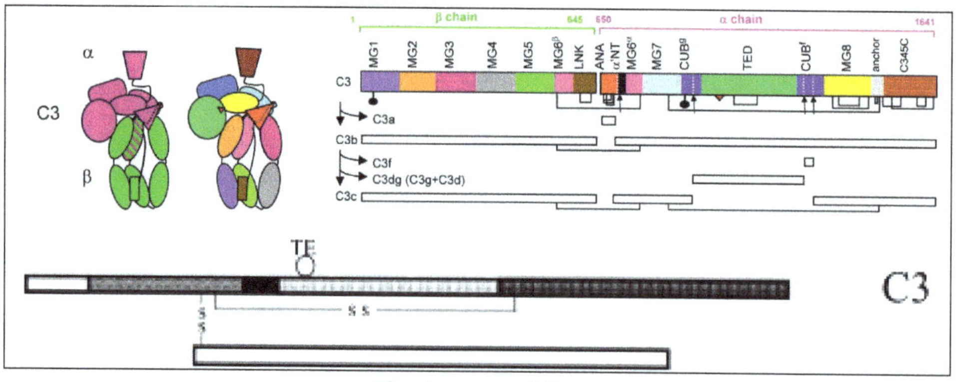

The structure of C3.

Alternative Pathway

C3 actually also breaks down spontaneously in plasma and as a result C3b is generated in normal serum at small concentrations all the time. C3b can combine with another protein to form alternative pathway C3 convertase. Thus, through this route C3b can form in adequate amount. A protein called factor H which can destroy C3b very fast. For alternate pathways to operate factor H activity has to be controlled. Factor H is a glycoprotein that can bind to αchain of C3b. This makes α-chain of C3b susceptible to proteolytic attack by factor I.

$$\text{Factor H- C3} \xrightarrow{\text{factor I}} \text{Inactive iC3b} \xrightarrow{\text{protease}} \text{C3c} + \text{C3d}$$

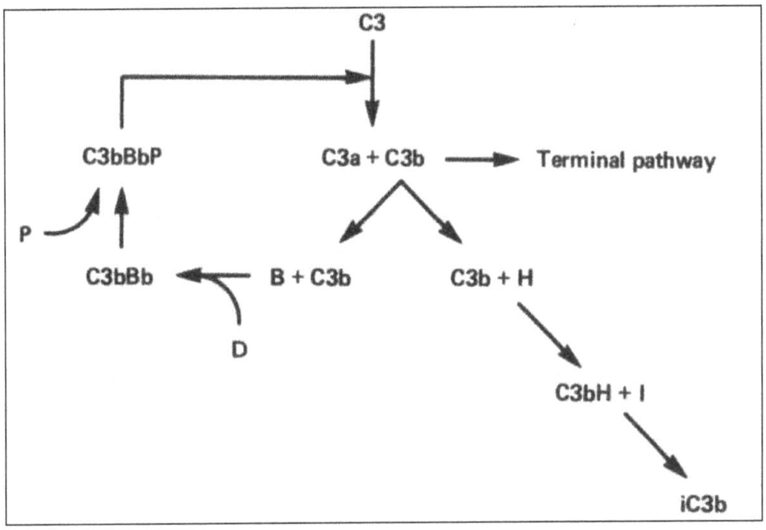

The alternate pathway of complement activation.

Factor H requires surface (on the antigen) rich in sialic acid to be able to bind C3b. For example, if some bacterial surface do not have enough sialic acid, factor H cannot bind to C3b. In that case C3b is able to bind to factor B. This leads to progressive activation of the complement pathway by positive feedback loop.

Factor B

Factor B is a large protein of 100 kDa molecular weight. Its binding to C3b is Mg++ dependent C3bB has weak C3 convertase activity. It can however acquire potent C3 convertase activity. Upon cleavage by factor D.

$$\text{B} \xrightarrow{\text{factor D}} \text{Ba(20 kda)} + \text{Bb (80Da)}$$

$$\text{C3b} + \text{Bb} \longrightarrow \text{c3bBb}$$

C3bBb is a potent C3 convertase. Factor D, is a serine esterase which actually cleaves B only when it is bound to C3b. So, C3b supply regulates this process. C3bBb has a very short half life of 5 minutes. Its half life is extended to 30 minutes if it binds to another protein called factor P or properdine. Factor P is a γ-globuln of 220 kDa molecular weight and can bind to C3bBb reversibly. Hence sometime alternative complement pathway has been referred to properdine pathway by some people.

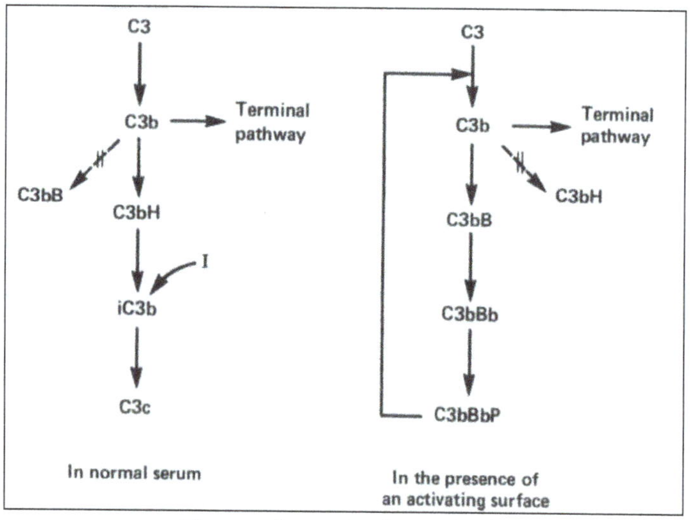

The two alternative fates of C3b.

The alternative pathway requires low level breakdown of C3. This cleavage generates a thioester bond in C3b which is reactive enough to bind to many molecule or cells. If C3b binds to a self-cell, there are regulating molecules which can inactivate it. If it binds to the invading microbe, Bb binds to this bound C3b. The alternative complement pathway can protect from infection unaided by lymphocytes. Hence, alternative complement pathway is older pathway in terms of evolution. It may be recollected that antibody formation takes some time to form. Hence during this period alternative pathway is the first line of defense. Classical pathway in fact evolved much later after the evolution of humoral immunity.

Complement Pathway

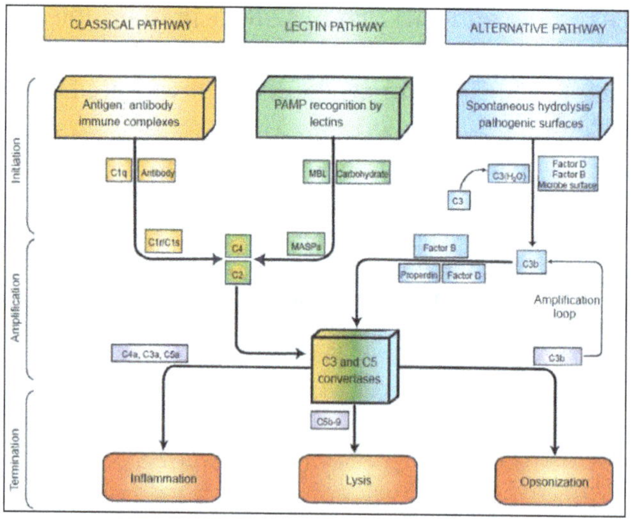

Figure shows the complement pathway. Complement can be activated through three pathways: classical, lectin, and alternative. The classical pathway is activated when C1q binds to antibody attached to antigen, activating C1r and C1s, which cleave C4 and C2. The lectin pathway is activated when mannose-binding lectin (MBL) encounters conserved pathogenic carbohydrate motifs, activating the MBL-associated serine proteases (MASPs) and again cleaving C4 and C2. C4 and C2 cleavage products form the classical and lectin pathway C3 convertase, C4bC2a, which cleaves C3 into C3b and C3a. A second molecule of C3b can associate with C4bC2a to form the C5 convertase of the classical and lectin pathways, C4bC2aC3b. The alternative pathway (AP) is activated when C3 undergoes spontaneous hydrolysis and forms the initial AP C3 convertase, $C3(H_2O)Bb$, in the presence of Factors B and D, leading to additional C3 cleavage and eventual formation of the AP C3 convertase (C3bBb) and AP C5 convertase (C3bBbC3b). Properdin facilitates AP activation by stabilizing AP convertases. All three pathways culminate in the formation of the convertases, which in turn generate the major effectors of the complement system: anaphylatoxins (C4a/C3a/C5a), the membrane attack complex (MAC), and opsonins (e.g., C3b). Anaphylatoxins are potent proinflammatory molecules derived from the cleavage of C4, C3, and C5. The MAC is a terminal assembly of complement components C5b through C9, which can directly lyse targeted surfaces. C3b induces phagocytosis of opsonized targets and also serves to amplify complement activation through the AP.

The classical complement pathway is the main pathway by which the complement system occurs. It is comprised of a cascade of many steps with complement proteins cleaving one another in a sequential order:

- The antibody binds to an antigen on the surface of a pathogen, activating the C1 complement protein.
- C1 acts a protease and cleaves C2 and C4 to form C4b2b.
- C42b converts C3 into C3a and C3b, which forms a C5 convertase.
- C5 convertase cleaves C5 into C5a and C5b.
- C5b forms a complex with C6, C7, and then C8, and C9, which becomes the membrane attack complex that lyses the pathogen.

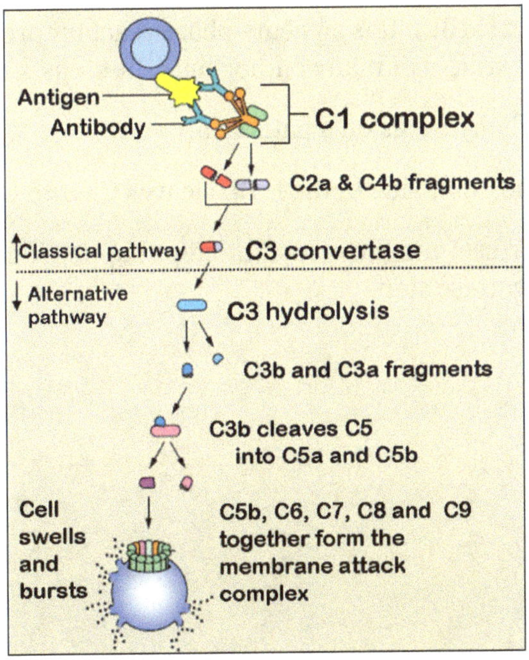

The Classical and Alternative Complement Pathways: The classical and alternative complement pathways start off differently, but end in the same cascade of complement proteins that combine to form a membrane attack complex.

Note that C5a has a number of other functions in the immune system, such as causing vasodilation during inflammation and stimulating neutrophil chemotaxis. Additionally, the body's cells express a glycoprotein called decay accelerating factor, which decays C3 and C5 convertase on the body's cells. This factor prevents membrane attack complexes from forming on the body's cells under normal conditions.

The Alternative Complement Pathway

The alternative pathway may be a leftover evolutionary precursor to the classical pathway. Unlike the classical pathway, the alternative pathway is generally activated by microbial inflammatory mediators instead of antibodies. For example, lipopolysaccharide, the toxin of gram-negative bacteria, may activate this pathway. The steps for the alternative pathway are:

- The pathogenic antigen (such as LPS) activates C3 so it creates a C3B complex.
- Factor D cleaves the C3B complex so that C3bBb is created.
- C3bBb is a C3 convertase, which converts more C3 into C3a and C3b.
- Similarly to the classical pathway, C3b forms a C42b complex, and the rest of the steps are essentially the same as the classical pathway, ending with C5b forming a membrane attack complex with C6, C7, C8, and C9.

Lectin Pathway

The lectin pathway is not caused by antibody binding, but by a carbohydrate -binding-protein called mannan-binding-lectin (MBL). It is an acute phase reactant produced in the liver and binds to the carbohydrates on the surfaces of many pathogens. The steps for the lectin pathway are:

- MBL binds to the carbohydrates on a pathogen.
- Protease bound on the other side of the MBL cleaves C4 into C4a and C4b.
- C4b creates C3 convertase, and the rest of the steps happen identically to the classical pathway from the C3 convertase step.

Chapter 5

Cytokines and Interferons

Cytokines are a large group of proteins, peptides and glycoproteins secreted by the cells of the immune system. They play the role of a signaling molecule, synchronizing the responses of the immune system. Cytokine nomenclature, cytokine receptor, cytokine regulation, signal transduction and pathways, interferon, etc. are a few other terms which are discussed in detail in this chapter.

Cytokines

Cytokines are a broad and loose category of small proteins (~5–20 kDa) that are important in cell signaling. Their release has an effect on the behavior of cells around them. It can be said that cytokines are involved in autocrine signalling, paracrine signalling and endocrine signalling as immunomodulating agents. Their definite distinction from hormones is still part of ongoing research. Cytokines include chemokines, interferons, interleukins, lymphokines, and tumour necrosis factors but generally not hormones or growth factors (despite some overlap in the terminology). Cytokines are produced by a broad range of cells, including immune cells like macrophages, B lymphocytes, T lymphocytes and mast cells, as well as endothelial cells, fibroblasts, and various stromal cells; a given cytokine may be produced by more than one type of cell.

They act through receptors, and are especially important in the immune system; cytokines modulate the balance between humoral and cell-based immune responses, and they regulate the maturation, growth, and responsiveness of particular cell populations. Some cytokines enhance or inhibit the action of other cytokines in complex ways.

They are different from hormones, which are also important cell signaling molecules, in that hormones circulate in less variable concentrations and hormones tend to be made by specific kinds of cells.

They are important in health and disease, specifically in host responses to infection, immune responses, inflammation, trauma, sepsis, cancer, and reproduction.

Discovery of Cytokines

Interferon-alpha, an interferon type I, was identified in 1957 as a protein that interfered with viral replication. The activity of interferon-gamma (the sole member of the interferon type II class) was described in 1965; this was the first identified lymphocyte-derived mediator. Macrophage migration inhibitory factor (MIF) was identified simultaneously in 1966 by John David and Barry Bloom.

In 1969 Dudley Dumonde proposed the term "lymphokine" to describe proteins secreted from lymphocytes and later, proteins derived from macrophages and monocytes in culture were called "monokines". As scientists learned more, it was understood that these proteins and others were part of a broader class of proteins involved in self-defense, and should be called "cytokines".

Difference from Hormones

Classic hormones circulate in nanomolar (10^{-9} M) concentrations that usually vary by less than one order of magnitude. In contrast, some cytokines (such as IL-6) circulate in picomolar (10^{-12} M) concentrations that can increase up to 1,000-fold during trauma or infection. The widespread distribution of cellular sources for cytokines may be a feature that differentiates them from hormones. Virtually all nucleated cells, but especially endo/epithelial cells and resident macrophages (many near the interface with the external environment) are potent producers of IL-1, IL-6, and TNF-α. In contrast, classic hormones, such as insulin, are secreted from discrete glands (e.g., the pancreas). As of 2008, the current terminology refers to cytokines as immunomodulating agents. However, more research is needed in this area of defining cytokines and hormones.

Part of the difficulty with distinguishing cytokines from hormones is that some of the immunomodulating effects of cytokines are systemic rather than local. For instance, to use hormone terminology, the action of cytokines may be autocrine or paracrine in chemotaxis or chemokinesis and endocrine as a pyrogen. Further, as molecules, cytokines are not limited to their immunomodulatory role.

Nomenclature

Cytokines have been classed as lymphokines, interleukins, and chemokines, based on their presumed function, cell of secretion, or target of action. Because cytokines are characterised by considerable redundancy and pleiotropism, such distinctions, allowing for exceptions, are obsolete.

- The term *interleukin* was initially used by researchers for those cytokines whose presumed targets are principally leukocytes. It is now used largely for designation of newer cytokine molecules and bears little relation to their presumed function. The vast majority of these are produced by T-helper cells.
- *Lymphokines*, produced by lymphocytes.
- *Monokines*, produced exclusively by monocytes.
- *Interferons*, involved in antiviral responses.

- *Colony stimulating factors*, support the growth of cells in semisolid media.
- *Chemokines* mediate chemoattraction (chemotaxis) between cells.

Classification

Structural

Structural homogeneity has been able to partially distinguish between cytokines that do not demonstrate a considerable degree of redundancy so that they can be classified into four types:

- The four-α-helix bundle family: member cytokines have three-dimensional structures with four bundles of α-helices. This family, in turn, is divided into three sub-families:
    - the IL-2 subfamily.
    - the interferon (IFN) subfamily.
    - the IL-10 subfamily.
        - The first of these three, the IL-2 subfamily, is the largest. It contains several non-immunological cytokines including erythropoietin (EPO) and thrombopoietin (TPO). Furthermore, four-α-helix bundle cytokines can be grouped into *long-chain* and *short-chain* cytokines.
- the IL-1 family, which primarily includes IL-1 and IL-18.
- the IL-17 family, which has yet to be completely characterized, though member cytokines have a specific effect in promoting proliferation of T-cells that cause cytotoxic effects.
- the cysteine-knot cytokines include members of the transforming-growth-factor-beta superfamily, including TGF-β1, TGF-β2 and TGF-β3.

Functional

A classification that proves more useful in clinical and experimental practice outside of structural biology divides immunological cytokines into those that enhance cellular immune responses, type 1 (TNFα, IFN-γ, etc.), and type 2 (TGF-β, IL-4, IL-10, IL-13, etc.), which favor antibody responses.

A key focus of interest has been that cytokines in one of these two sub-sets tend to inhibit the effects of those in the other. Dysregulation of this tendency is under intensive study for its possible role in the pathogenesis of autoimmune disorders.

Several inflammatory cytokines are induced by oxidative stress. The fact that cytokines themselves trigger the release of other cytokines and also lead to increased oxidative stress makes them important in chronic inflammation, as well as other immunoresponses, such as fever and acute phase proteins of the liver (IL-1,6,12, IFN-a).

Cytokines also play a role in anti-inflammatory pathways and are a possible therapeutic treatment for pathological pain from inflammation or peripheral nerve injury. There are both pro-inflammatory and anti-inflammatory cytokines that regulate this pathway.

Receptors

In recent years, the cytokine receptors have come to demand the attention of more investigators than cytokines themselves, partly because of their remarkable characteristics, and partly because a deficiency of cytokine receptors has now been directly linked to certain debilitating immunodeficiency states. In this regard, and also because the redundancy and pleomorphism of cytokines are, in fact, a consequence of their homologous receptors, many authorities think that a classification of cytokine receptors would be more clinically and experimentally useful.

A classification of cytokine receptors based on their three-dimensional structure has, therefore, been attempted. Such a classification, though seemingly cumbersome, provides several unique perspectives for attractive pharmacotherapeutic targets.

- Immunoglobulin (Ig) superfamily, which are ubiquitously present throughout several cells and tissues of the vertebrate body, and share structural homology with immunoglobulins (antibodies), cell adhesion molecules, and even some cytokines. Examples: IL-1 receptor types.
- Hemopoietic Growth Factor (type 1) family, whose members have certain conserved motifs in their extracellular amino-acid domain. The IL-2 receptor belongs to this chain, whose γ-chain (common to several other cytokines) deficiency is directly responsible for the x-linked form of Severe Combined Immunodeficiency (X-SCID).
- Interferon (type 2) family, whose members are receptors for IFN β and γ.
- Tumor necrosis factors (TNF) (type 3) family, whose members share a cysteine-rich common extracellular binding domain, and includes several other non-cytokine ligands like CD40, CD27 and CD30, besides the ligands on which the family is named (TNF).
- Seven transmembrane helix family, the ubiquitous receptor type of the animal kingdom. All G protein-coupled receptors (for hormones and neurotransmitters) belong to this family. Chemokine receptors, two of which act as binding proteins for HIV (CD4 and CCR5), also belong to this family.
- Interleukin-17 receptor (IL-17R) family, which shows little homology with any other cytokine receptor family. Structural motifs conserved between members

of this family include: an extracellular fibronectin III-like domain, a transmembrane domain and a cytoplasmic SERIF domain. The known members of this family are as follows: IL-17RA, IL-17RB, IL-17RC, IL17RD and IL-17RE.

Cellular Effects

Each cytokine has a matching cell-surface receptor. Subsequent cascades of intracellular signalling then alter cell functions. This may include the upregulation and/or downregulation of several genes and their transcription factors, resulting in the production of other cytokines, an increase in the number of surface receptors for other molecules, or the suppression of their own effect by feedback inhibition.

The effect of a particular cytokine on a given cell depends on the cytokine, its extracellular abundance, the presence and abundance of the complementary receptor on the cell surface, and downstream signals activated by receptor binding; these last two factors can vary by cell type. Cytokines are characterized by considerable "redundancy", in that many cytokines appear to share similar functions.

It seems to be a paradox that cytokines binding to antibodies have a stronger immune effect than the cytokine alone. This may lead to lower therapeutic doses.

Said et al. showed that inflammatory cytokines cause an IL-10-dependent inhibition of T-cell expansion and function by up-regulating PD-1 levels on monocytes which leads to IL-10 production by monocytes after binding of PD-1 by PD-L.

Adverse reactions to cytokines are characterized by local inflammation and/or ulceration at the injection sites. Occasionally such reactions are seen with more widespread papular eruptions.

Roles of Endogenous Cytokines in Health and Disease

Cytokines are often involved in several developmental processes during embryogenesis.

Cytokines are crucial for fighting off infections and in other immune responses. However, they can become dysregulated and pathological in inflammation, trauma, and sepsis.

Adverse effects of cytokines have been linked to many disease states and conditions ranging from schizophrenia, major depression and Alzheimer's disease to cancer. Normal tissue integrity is preserved by feedback interactions between diverse cell types mediated by adhesion molecules and secreted cytokines; disruption of normal feedback mechanisms in cancer threatens tissue integrity. Over-secretion of cytokines can trigger a dangerous syndrome known as a cytokine storm; this may have been the cause of severe adverse events during a clinical trial of TGN1412. Cytokine storms are suspected to be the main cause of death in the 1918 "Spanish Flu" pandemic. Deaths were weighted more heavily towards people with healthy immune systems, due to its ability to produce stronger immune responses, likely increasing cytokine levels. Another important example of cytokine storm is seen in acute pancreatitis. Cytokines are integral and implicated in all angles of the cascade resulting in the systemic inflammatory response syndrome and multi organ failure associated with this intra-abdominal catastrophe.

Medical use as Drugs

Some cytokines have been developed into protein therapeutics using recombinant DNA technology. Recombinant cytokines being used as drugs as of 2014 include:

- Bone morphogenetic protein (BMP), used to treat bone-related conditions.
- Erythropoietin (EPO), used to treat anemia.
- Granulocyte colony-stimulating factor (G-CSF), used to treat neutropenia in cancer patients.
- Granulocyte macrophage colony-stimulating factor (GM-CSF), used to treat neutropenia and fungal infections in cancer patients.
- Interferon alfa, used to treat hepatitis C and multiple scelerosis.
- Interferon beta, used to treat multiple scelerosis.
- Interleukin 2 (IL-2), used to treat cancer.
- Interleukin 11 (IL-11), used to treat thrombocytopenia in cancer patients.
- Interferon gamma is used to treat chronic granulomatous disease and osteopetrosis.

Research into Diagnostic Use of Measured Levels

Plasma levels of various cytokines may give information on the presence, or even predictive value of inflammatory processes involved in autoimmune diseases such as rheumatoid arthritis, as well as immunomodulatory effects of foods or drugs. In addition, elevated levels of IL-7, an important cytokine involved in T cell homeostasis, have been detected in the plasma of HIV-infected patients.

Features of Cytokines

- Cytokines show redundancy i.e. different cytokines may have similar function, e.g. IL-1 and IL-6 both induce fever by acting on brain.
- They are secreted in multiple numbers, e.g. macrophages secrete more than five interleukins and tumor necrosis factor.

- Cytokines produce their effect on many cell types i.e. they are pleiotropic.
- Cytokines need extreme regulation and are toxic in high doses.

Cytokines Nomenclature

Most of the cytokines are named according to the Interleukin nomenclature subcommittee of the international union of immunological societies. Although the definition of cytokines is quite broad, but together they can be classified as lymphokines, interleukins, interferons , chemokines etc. depending on their function, cell of secretion, or target of action. The name interleukin was coined initially for those cytokines that mediate signaling between lymphocytes and leukocytes. In other words leukocytes were thought to be the principal target for interleukins but now this definition is no longer in function and the name interleukin is given to any new cytokine discovered routinely. Cytokines are produced mainly in response to viral infection or in response to immune attack. Interferons are the cytokines that interfere with viral RNA and protein synthesis to prevent viral replication.

Types of interferon – Interferons are mainly divided into two types.

Type I Interferons --- Interferon-α (IFN- α) and interferon-β (IFN-β).

Type II interferon --- Interferon-γ.

Tumor necrosis factors (TNFs) can kill tumor cells and colony stimulating factors assist in regulation of stem cell activities. Chemokines play role in leukocyte recruitment and circulation.

Table: Different cytokines and their examples.

Cytokine types	Examples
Lymphokines	MAF (Macrophage activating factor)
Interleukins	IL-7, IL-13, IL1-7
Interferons	IFN-α, IFN –β, IFN-γ, IFN-ω, IFN -
Tumor necrosis factors	TNF- α (Cachectin), TNF – β (lymphotoxin)
Colony stimulating factors	GSF (Granulocyte colony stimulating factor)
Polypeptide growth factors	EGF (Epidermal growth factor)
α-Chemokines	PF-4 (Platelet factor 4)
β-Chemokines	MCP-1 (Monocyte chemoattractant protein 1)
Transforming growth factors	TGF- α, TGF - β
Stress proteins	Heat shock proteins
RANTES	

Lymphokine

Lymphokines are a subset of cytokines that are produced by a type of immune cell known as a lymphocyte. They are protein mediators typically produced by T cells to direct the immune system response by signaling between its cells. Lymphokines have many roles, including the attraction of other immune cells, including macrophages and other lymphocytes, to an infected site and their subsequent activation to prepare them to mount an immune response. Circulating lymphocytes can detect a very small concentration of lymphokine and then move up the concentration gradient towards where the immune response is required. Lymphokines aid B cells to produce antibodies.

Important lymphokines secreted by the T helper cell include:

- Interleukin 2.
- Interleukin 3.
- Interleukin 4.
- Interleukin 5.
- Interleukin 6.
- Granulocyte-macrophage colony-stimulating factor.
- Interferon-gamma.

Chemokine

Chemokines (movement) are a family of small cytokines, or signaling proteins secreted by cells. Their name is derived from their ability to induce directed chemotaxis in nearby responsive cells; they are chemotactic cytokines.

Cytokine proteins are classified as chemokines according to behavior and structural characteristics. In addition to being known for mediating chemotaxis, chemokines are all approximately 8-10 kilodaltons in mass and have four cysteine residues in conserved locations that are key to forming their 3-dimensional shape.

These proteins have historically been known under several other names including the *SIS family of cytokines*, *SIG family of cytokines*, *SCY family of cytokines*, *Platelet factor-4 superfamily* or *intercrines*. Some chemokines are considered pro-inflammatory and can be induced during an immune response to recruit cells of the immune system to a site of infection, while others are considered homeostatic and are involved in controlling the migration of cells during normal processes of tissue maintenance or development. Chemokines are found in all vertebrates, some viruses and some bacteria, but none have been described for other invertebrates.

Chemokines have been classified into four main subfamilies: CXC, CC, CX3C and XC.

All of these proteins exert their biological effects by interacting with G protein-linked transmembrane receptors called chemokine receptors, that are selectively found on the surfaces of their target cells.

Function

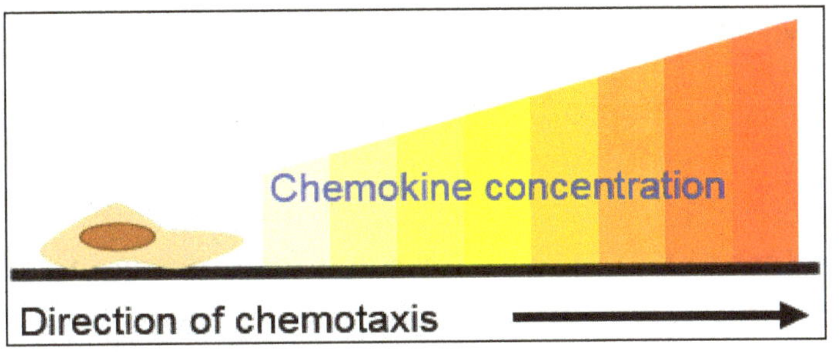

Chemokines released by infected or damaged cells form a concentration gradient. Attracted cells move through the gradient towards the higher concentration of chemokine.

The major role of chemokines is to act as a chemoattractant to guide the migration of cells. Cells that are attracted by chemokines follow a signal of increasing chemokine concentration towards the source of the chemokine. Some chemokines control cells of the immune system during processes of immune surveillance, such as directing lymphocytes to the lymph nodes so they can screen for invasion of pathogens by interacting with antigen-presenting cells residing in these tissues. These are known as homeostatic chemokines and are produced and secreted without any need to stimulate their source cell(s). Some chemokines have roles in development; they promote angiogenesis (the growth of new blood vessels), or guide cells to tissues that provide specific signals critical for cellular maturation. Other chemokines are inflammatory and are released from a wide variety of cells in response to bacterial infection, viruses and agents that cause physical damage such as silica or the urate crystals that occur in gout. Their release is often stimulated by pro-inflammatory cytokines such as interleukin 1. Inflammatory chemokines function mainly as chemoattractants for leukocytes, recruiting monocytes, neutrophils and other effector cells from the blood to sites of infection or tissue damage. Certain inflammatory chemokines activate cells to initiate an immune response or promote wound healing. They are released by many different cell types and serve to guide cells of both innate immune system and adaptive immune system.

Types by function

Chemokines are functionally divided into two groups:

- Homeostatic: are constitutively produced in certain tissues and are responsible for basal leukocyte migration. These include: CCL14, CCL19, CCL20, CCL21, CCL25, CCL27, CXCL12 and CXCL13. This classification is not strict; for example, CCL20 can act also as pro-inflammatory chemokine.
- Inflammatory: these are formed under pathological conditions (on pro-inflammatory stimuli, such as IL-1, TNF-alpha, LPS, or viruses) and actively participate

in the inflammatory response attracting immune cells to the site of inflammation. Examples are: CXCL-8, CCL2, CCL3, CCL4, CCL5, CCL11, CXCL10.

Homing

The main function of chemokines is to manage the migration of leukocytes (homing) in the respective anatomical locations in inflammatory and homeostatic processes.

Basal: homeostatic chemokines are basal produced in the thymus and lymphoid tissues. Their homeostatic function in homing is best exemplified by the chemokines CCL19 and CCL21 (expressed within lymph nodes and on lymphatic endothelial cells) and their receptor CCR7 (expressed on cells destined for homing in cells to these organs). Using these ligands is possible routing antigen-presenting cells (APC) to lymph nodes during the adaptive immune response. Among other homeostatic chemokine receptors include: CCR9, CCR10, and CXCR5, which are important as part of the cell addresses for tissue-specific homing of leukocytes. CCR9 supports the migration of leukocytes into the intestine, CCR10 to the skin and CXCR5 supports the migration of B-cell to follicles of lymph nodes. As well CXCL12 (SDF-1) constitutively produced in the bone marrow promotes proliferation of progenitor B cells in the bone marrow microenvironment.

Inflammatory: inflammatory chemokines are produced in high concentrations during infection or injury and determine the migration of inflammatory leukocytes into the damaged area. Typical inflammatory chemokines include: CCL2, CCL3 and CCL5, CXCL1, CXCL2 and CXCL8. A typical example is CXCL-8, which chemoatractants for neutrophils. In contrast to the homeostatic chemokine receptors, there is significantly promiscuous (redundancy) binding receptor and inflammatory chemokines. This often complicated research on receptor-specific therapeutics in this area.

Types by cell attracted

- Monocytes / macrophages: the key chemokines that attract these cells to the site of inflammation include: CCL2, CCL3, CCL5, CCL7, CCL8, CCL13, CCL17 and CCL22.
- T-lymphocytes: the four key chemokines that are involved in the recruitment of T lymphocytes to the site of inflammation are: CCL2, CCL1, CCL22 and CCL17.
- Mast cells: on their surface express several receptors for chemokines: CCR1, CCR2, CCR3, CCR4, CCR5, CXCR2, and CXCR4. Ligands of these receptors CCL2 and CCL5 play an important role in mast cell recruitment and activation in the lung.There is also evidence that CXCL8 might be inhibitory of mast cell.
- Eosinophils: the migration of eosinophils into various tissues involved several

chemokines of CC family: CCL11, CCL24, CCL26, CCL5, CCL7, CCL13, and CCL3. Chemokines CCL11 (eotaxin) and CCL5 (RANTES) acts through a specific receptor CCR3 on the surface of eosinophils, and eotaxin plays an essential role in the initial recruitment of eosinophils into the lesion.

- Neutrophils: are regulated primarily by CXC chemokines. An example CXCL8 (IL-8) is chemoattractant for neutrophils and also activating their metabolic and degranulation.

Structural Characteristics

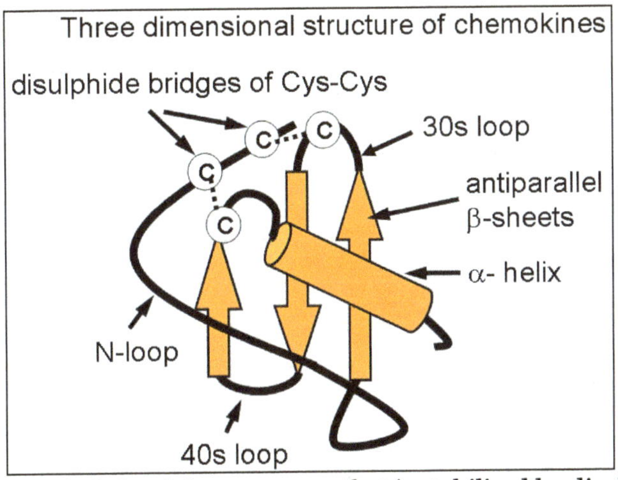

All chemokines share a typical *Greek key* structure that is stabilised by disulfide bonds between conserved cysteine residues.

Proteins are classified into the chemokine family based on their structural characteristics, not just their ability to attract cells. All chemokines are small, with a molecular mass of between 8 and 10 kDa. They are approximately 20-50% identical to each other; that is, they share gene sequence and amino acid sequence homology. They all also possess conserved amino acids that are important for creating their 3-dimensional or tertiary structure, such as (in most cases) four cysteines that interact with each other in pairs to create a Greek key shape that is a characteristic of chemokines. Intramolecular disulfide bonds typically join the first to third, and the second to fourth cysteine residues, numbered as they appear in the protein sequence of the chemokine. Typical chemokine proteins are produced as pro-peptides, beginning with a signal peptide of approximately 20 amino acids that gets cleaved from the active (mature) portion of the molecule during the process of its secretion from the cell. The first two cysteines, in a chemokine, are situated close together near the N-terminal end of the mature protein, with the third cysteine residing in the centre of the molecule and the fourth close to the C-terminal end. A loop of approximately ten amino acids follows the first two cysteines and is known as the *N-loop*. This is followed by a single-turn helix, called a 3_{10}-helix, three β-strands and a C-terminal α-helix. These helices and strands are connected by turns called *30s*, *40s* and *50s* loops; the third and fourth cysteines are located in the 30s and 50s loops.

Types by Structure

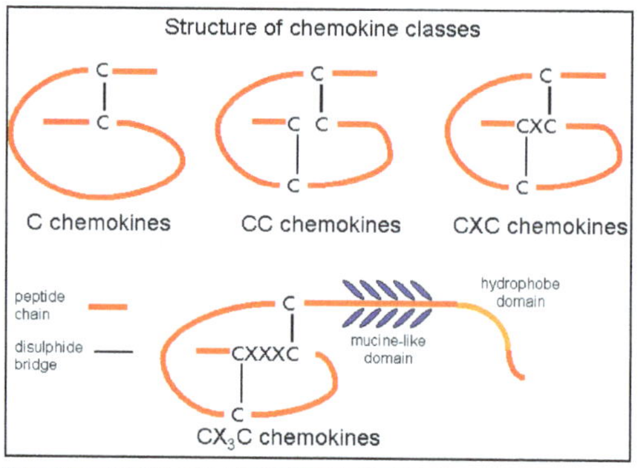

CC chemokines				
Name	**Gene**	**Other name(s)**	**Receptor**	**Uniprot**
CCL1	Scya1	I-309, TCA-3	CCR8	
CCL2	Scya2	MCP-1	CCR2	P13500
CCL3	Scya3	MIP-1a	CCR1	P10147
CCL4	Scya4	MIP-1β	CCR1, CCR5	P13236
CCL5	Scya5	RANTES	CCR5	P13501
CCL6	Scya6	C10, MRP-2	CCR1	P27784
CCL7	Scya7	MARC, MCP-3	CCR2	P80098
CCL8	Scya8	MCP-2	CCR1, CCR2B, CCR5	P80075
CCL9/CCL10	Scya9	MRP-2, CCF18, MIP-1?	CCR1	P51670
CCL11	Scya11	Eotaxin	CCR2, CCR3, CCR5	P51671
CCL12	Scya12	MCP-5		Q62401
CCL13	Scya13	MCP-4, NCC-1, Ckβ10	CCR2, CCR3, CCR5	Q99616
CCL14	Scya14	HCC-1, MCIF, Ckβ1, NCC-2, CCL	CCR1	Q16627
CCL15	Scya15	Leukotactin-1, MIP-5, HCC-2, NCC-3	CCR1, CCR3	Q16663
CCL16	Scya16	LEC, NCC-4, LMC, Ckβ12	CCR1, CCR2, CCR5, CCR8	O15467
CCL17	Scya17	TARC, dendrokine, ABCD-2	CCR4	Q92583
CCL18	Scya18	PARC, DC-CK1, AMAC-1, Ckβ7, MIP-4		P55774
CCL19	Scya19	ELC, Exodus-3, Ckβ11	CCR7	Q99731
CCL20	Scya20	LARC, Exodus-1, Ckβ4	CCR6	P78556
CCL21	Scya21	SLC, 6Ckine, Exodus-2, Ckβ9, TCA-4	CCR7	O00585
CCL22	Scya22	MDC, DC/β-CK	CCR4	O00626
CCL23	Scya23	MPIF-1, Ckβ8, MIP-3, MPIF-1	CCR1	P55773
CCL24	Scya24	Eotaxin-2, MPIF-2, Ckβ6	CCR3	O00175
CCL25	Scya25	TECK, Ckβ15	CCR9	O15444
CCL26	Scya26	Eotaxin-3, MIP-4a, IMAC, TSC-1	CCR3	Q9Y258
CCL27	Scya27	CTACK, ILC, Eskine, PESKY, skinkine	CCR10	Q9Y4X3
CCL28	Scya28	MEC	CCR3, CCR10	Q9NRJ3
CXCL1	Scyb1	Gro-a, GRO1, NAP-3, KC	CXCR2	P09341

CXC chemokines				
Name	**Gene**	**Other name(s)**	**Receptor**	**Uniprot**
CXCL2	Scyb2	Gro-β, GRO2, MIP-2a	CXCR2	P19875
CXCL3	Scyb3	Gro-?, GRO3, MIP-2β	CXCR2	P19876
CXCL4	Scyb4	PF-4	CXCR3B	P02776
CXCL5	Scyb5	ENA-78	CXCR2	P42830
CXCL6	Scyb6	GCP-2	CXCR1, CXCR2	P80162
CXCL7	Scyb7	NAP-2, CTAPIII, β-Ta, PEP		P02775
CXCL8	Scyb8	IL-8, NAP-1, MDNCF, GCP-1	CXCR1, CXCR2	P10145
CXCL9	Scyb9	MIG, CRG-10	CXCR3	Q07325
CXCL10	Scyb10	IP-10, CRG-2	CXCR3	P02778
CXCL11	Scyb11	I-TAC, β-R1, IP-9	CXCR3, CXCR7	014625
CXCL12	Scyb12	SDF-1, PBSF	CXCR4, CXCR7	P48061
CXCL13	Scyb13	BCA-1, BLC	CXCR5	043927
CXCL14	Scyb14	BRAK, bolekine		095715
CXCL15	Scyb15	Lungkine, WECHE		Q9WVL7
CXCL16	Scyb16	SRPSOX	CXCR6	Q9H2A7
CXCL17	VCC-1	DMC, VCC-1		Q6UXB2
C chemokines				
Name	**Gene**	**Other name(s)**	**Receptor**	**Uniprot**
XCL1	Scyc1	Lymphotactin a, SCM-1a, ATAC	XCR1	P47992
XCL2	Scyc2	Lymphotactin β, SCM-1β	XCR1	Q9UBD3
CX3C chemokines				
Name	**Gene**	**Other name(s)**	**Receptor**	**Uniprot**
CX3CL1	Scyd1	Fractalkine, Neurotactin, ABCD-3	CX3CR1	P78423

Members of the chemokine family are divided into four groups depending on the spacing of their first two cysteine residues. Thus the nomenclature for chemokines is, e.g.: CCL1 for the ligand 1 of the CC-family of chemokines, and CCR1 for its respective receptor.

CC Chemokines

The CC chemokine (or β-chemokine) proteins have two adjacent cysteines (amino acids), near their amino terminus. There have been at least 27 distinct members of this subgroup reported for mammals, called CC chemokine ligands (CCL)-1 to -28; CCL10 is the same as CCL9. Chemokines of this subfamily usually contain four cysteines (C4-CC chemokines), but a small number of CC chemokines possess six cysteines (C6-CC chemokines). C6-CC chemokines include CCL1, CCL15, CCL21, CCL23 and CCL28. CC chemokines induce the migration of monocytes and other cell types such as NK cells and dendritic cells.

Examples of CC chemokine include monocyte chemoattractant protein-1 (MCP-1 or CCL2) which induces monocytes to leave the bloodstream and enter the surrounding tissue to become tissue macrophages.

CCL5 (or RANTES) attracts cells such as T cells, eosinophils and basophils that express the receptor CCR5.

Increased CCL11 levels in blood plasma are associated with aging (and reduced neurogenesis) in mice and humans.

CXC Chemokines

The two N-terminal cysteines of CXC chemokines (or α-chemokines) are separated by one amino acid, represented in this name with an "X". There have been 17 different CXC chemokines described in mammals, that are subdivided into two categories, those with a specific amino acid sequence (or motif) of glutamic acid-leucine-arginine (or ELR for short) immediately before the first cysteine of the CXC motif (ELR-positive), and those without an ELR motif (ELR-negative). ELR-positive CXC chemokines specifically induce the migration of neutrophils, and interact with chemokine receptors CXCR1 and CXCR2. An example of an ELR-positive CXC chemokine is interleukin-8 (IL-8), which induces neutrophils to leave the bloodstream and enter into the surrounding tissue. Other CXC chemokines that lack the ELR motif, such as CXCL13, tend to be chemoattractant for lymphocytes. CXC chemokines bind to CXC chemokine receptors, of which seven have been discovered to date, designated CXCR1-7.

C Chemokines

The third group of chemokines is known as the C chemokines (or γ chemokines), and is unlike all other chemokines in that it has only two cysteines; one N-terminal cysteine and one cysteine downstream. Two chemokines have been described for this subgroup and are called XCL1 (lymphotactin-α) and XCL2 (lymphotactin-β).

CX3C Chemokines

A fourth group has also been discovered and members have three amino acids between the two cysteines and is termed CX_3C chemokine (or d-chemokines). The only CX_3C chemokine discovered to date is called fractalkine (or CX_3CL1). It is both secreted and tethered to the surface of the cell that expresses it, thereby serving as both a chemoattractant and as an adhesion molecule.

Receptors

Chemokine receptors are G protein-coupled receptors containing 7 transmembrane domains that are found on the surface of leukocytes. Approximately 19 different chemokine receptors have been characterized to date, which are divided into four families depending on the type of chemokine they bind; CXCR that bind CXC chemokines, CCR that bind CC chemokines, CX3CR1 that binds the sole CX3C chemokine (CX3CL1), and XCR1 that binds the two XC chemokines (XCL1 and XCL2). They share many structural features; they are similar in size (with about 350 amino acids), have a short, acidic N-terminal end, seven helical transmembrane domains with three intracellular and three extracellular hydrophilic loops, and an intracellular C-terminus containing serine and threonine residues important for receptor regulation. The first two extracellular loops of chemokine receptors each has a conserved cysteine residue that allow formation of a disulfide bridge between these loops. G proteins are coupled to the C-terminal end of the chemokine receptor to allow intracellular signaling after receptor activation, while the N-terminal domain of the chemokine receptor determines ligand binding specificity.

Signal Transduction

Chemokine receptors associate with G-proteins to transmit cell signals following ligand binding. Activation of G proteins, by chemokine receptors, causes the subsequent activation of an enzyme known as phospholipase C (PLC). PLC cleaves a molecule called phosphatidylinositol (4,5)-bisphosphate (PIP2) into two second messenger molecules known as Inositol triphosphate (IP3) and diacylglycerol (DAG) that trigger intracellular signaling events; DAG activates another enzyme called protein kinase C (PKC), and IP3 triggers the release of calcium from intracellular stores. These events promote many signaling cascades (such as the MAP kinase pathway) that generate responses like chemotaxis, degranulation, release of superoxide anions and changes in the avidity of cell adhesion molecules called integrins within the cell harbouring the chemokine receptor.

Infection Control

The discovery that the β chemokines RANTES, MIP (macrophage inflammatory proteins) 1α and 1β (now known as CCL5, CCL3 and CCL4 respectively) suppress HIV-1 provided the initial connection and indicated that these molecules might control infection as part of immune responses in vivo, and that sustained delivery of such inhibitors have the capacity of long-term infection control. The association of chemokine production with antigen-induced proliferative responses, more favorable clinical status in HIV infection, as well as with an uninfected status in subjects at risk for infection suggests a positive role for these molecules in controlling the natural course of HIV infection.

Functions of Cytokines

Cytokines are crucial regulators of cells and hence control their growth, movement, evolution and differentiation. Being responsible for so many factors they are generated in response to varied stimuli. The most critical of these stimuli are antigen-antibody complexes acting through antibody receptors (FcRs), pathogen-associated molecular patterns (PAMPs) such as lipopolysaccharides acting through toll-like receptors (TLRs) and antigens acting through the B cell or T cell antigen receptors. Cytokines acts on different cells and in different ways. Cytokines may have an autocrine effect (Cytokines bind to receptors that produced them), paracrine effect (Cytokines bind to receptors on neighboring or cells which are very close) or endocrine effect (Cytokines flow throughout the body) while acting on the target cells. Cytokines also play role in apoptosis and control various stages of replication. They have the capability for being sensitive markers of chemically induced perturbations in function, but this claim is not accepted from the toxicological point of view. The reasons behind this being that cytokines are released locally with plasma measures being unreliable and they have short half-lives which need precise timing to find.

Cytokine structure

Cytokines have distinct structures but can be broadly divided into following.

Group I cytokines are involved in immune regulation --- includes four α -helices in bundles, Interferon subfamily and the IL-10 subfamily.

Group II cytokines are engaged in growth and regulation of cells, cell death and inflammation --- IL-1 family, TGF-β.

Group III cytokines are involved in inflammation--- Chemokines.

Group IV cytokines --- IL-12.

Table: Different cytokines and their source:

Cytokine	Principal Cell Source
Interleukin-1 (IL -1)	Macrophages, endothelial cells, some epithelial cells
Interleukin-6 (IL -6)	Macrophages, endothelial cells, T cells
Interleukin-10 (IL-10)	Macrophages, T cells (mainly regulatory T cells)
Interleukin-12 (IL-12)	Macrophages, dendritic cells
Interleukin-15 (IL-15)	Macrophages, others
Interleukin-18 (IL-18)	Macrophages
Interleukin-23 (IL-23)	Macrophages and dendritic cells
Interleukin-27 (IL-27)	Macrophages and dendritic cells
Chemokines	Macrophages, endothelial cells, T cells, fibroblasts, platelets
Type I interferons (IFN -α, IFN-β)	IFN-α: macrophages; IFN-β: fibroblasts
Tumor necrosis factor (TNF)	Macrophages, T cells

Cytokine Receptor

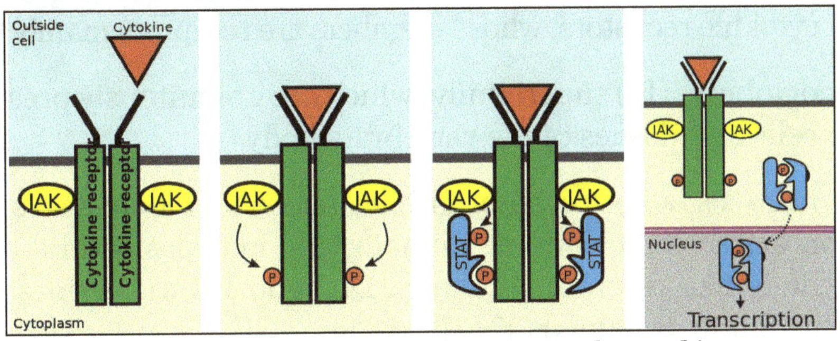

Key steps of the JAK-STAT pathway for type 1 and 2 cytokine receptors.

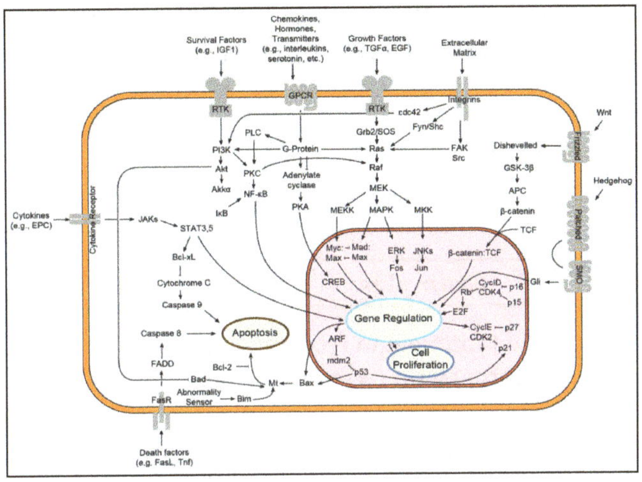

Signal transduction. (Cytokine receptor at center left.)

Cytokine receptors are receptors that bind cytokines.

In recent years, the cytokine receptors have come to demand the attention of more investigators than cytokines themselves, partly because of their remarkable characteristics, and partly because a deficiency of cytokine receptors has now been directly linked to certain debilitating immunodeficiency states. In this regard, and also because the redundancy and pleiotropy of cytokines are a consequence of their homologous receptors, many authorities are now of the opinion that a classification of cytokine receptors would be more clinically and experimentally useful.

Classification

A classification of cytokine receptors based on their three-dimensional structure has been attempted. (Such a classification, though seemingly cumbersome, provides several unique perspectives for attractive pharmacotherapeutic targets.)

- Type I cytokine receptors, whose members have certain conserved motifs in their extracellular amino-acid domain. The IL-2 receptor belongs to this chain, whose γ-chain (common to several other cytokines) deficiency is directly responsible for the x-linked form of Severe Combined Immunodeficiency (X-SCID).

- Type II cytokine receptors, whose members are receptors mainly for interferons.
- Immunoglobulin (Ig) superfamily, which are ubiquitously present throughout several cells and tissues of the vertebrate body.
- Tumor necrosis factor receptor family, whose members share a cysteine-rich common extracellular binding domain, and includes several other non-cytokine ligands like receptors, CD40, CD27 and CD30, besides the ligands on which the family is named (TNF).
- Chemokine receptors, two of which acting as binding proteins for HIV (CXCR4 and CCR5). They are G protein coupled receptors.
- TGF beta receptors.

Comparison

<table>
<tr><th>Type</th><th>Examples</th><th>Structure</th><th>Mechanism</th></tr>
<tr><td>type I cytokine receptor</td><td>• Type 1 interleukin receptors
• Erythropoietin receptor
• GM-CSF receptor
• G-CSF receptor
• growth hormone receptor
• prolactin receptor
• Oncostatin M receptor
• Leukemia inhibitory factor receptor</td><td rowspan="2">Certain conserved motifs in their extracellular amino-acid domain. Connected to Janus kinase (JAK) family of tyrosine kinases</td><td rowspan="2">JAK phosphorylate and activate downstream proteins involved in their signal transduction pathways</td></tr>
<tr><td>type II cytokine receptor</td><td>• Type II interleukin receptors
• interferon-alpha/beta receptor
• interferon-gamma receptor</td></tr>
<tr><td>Many members of the immunoglobulin superfamily</td><td>• Interleukin-1 receptor
• CSF1
• C-kit receptor
• Interleukin-18 receptor</td><td>Share structural homology with immunoglobulins (antibodies), cell adhesion molecules, and even some cytokines.</td><td></td></tr>
<tr><td>Tumor necrosis factor receptor family</td><td>• CD27
• CD30
• CD40
• CD120
• Lymphotoxin beta receptor</td><td>cysteine-rich common extracellular binding domain</td><td></td></tr>
</table>

chemokine receptors	• Interleukin-8 receptor • CCR1 • CXCR4 • MCAF receptor • NAP-2 receptor	Seven transmembrane helix	G protein-coupled
TGF beta receptors	• TGF beta receptor 1 • TGF beta receptor 2		

Solubility

Cytokine receptors may be both membrane-bound and soluble. *Soluble cytokine receptors* are extremely common regulators of cytokine function. Soluble cytokine receptors typically consist of the extracellular portions of membrane-bound receptors.

Cytokine Regulation

The three major ways for regulating the cytokine signaling are –

- By specific binding proteins.
- By changes in receptor expression.
- By cytokines that exert opposite effects.

Concept behind understanding the cytokine regulation is that as the single cell receives signals from numerous cytokine receptors, a unification of multiple signals is required to generate a coherent response.

Signal Transduction and Pathways

During the signal transduction the first step is the binding of cytokines to its receptors. Once the binding occurs, the receptor transmits a signal to the cell to alter its behavior. This phenomenon involving conversion of an extracellular signal into a series of intracellular events is called signal transduction. As Cell signaling needs to be precise and quick, enzyme cascade serve this function by the amplification of the responses rapidly. The important features of signal transduction include binding of cytokine to a cell surface receptor, stimulation of a transducer protein by the receptor, secondary activation of various enzymes, production of new transcription factors and ultimately gene stimulation leading to changed cell behavior.

Although there are many pathways of signal transduction but three pathways are of utmost significance to the immune system.

- The NF-kB pathway.
- The NF-AT pathway.
- The JAK/STAT pathway.

The NF-κβ Pathway

NF-κβ has its origins in the family of five transcription factors that play a crucial role in inflammation and immunity. There are more than 150 genes that get expressed on NF-κβ stimulation. There are three major pathways for NF-κβ activation. These are the classical pathway , alternative pathway and the third pathway that is stimulated by DNA damaging drugs and Ultraviolet light. The classical pathway involves proinflammatory signaling and gets activated by inflammatory cytokines, TLRs and antigen receptors. IKK (Iκβ) complex is the central regulator of NF-κβ and all the signals induced by various stimuli focus on the same complex IKK. IKK complex is rich with kinase activity and on activation IKK complex phosphorylates Iκβ. Consequently Iκβ dissociates from the NF-κβ and the newly formed Iκβ becomes ubiquinated and is destroyed by proteasomes. Dissociation between Iκβ and NF-κβ liberates NF-κβ and it enters the nucleus to activate κβ motif containing promoters on DNA. This stimulates many genes and cytokines IL-1β, IL-6, IL-18, IL-33, TNF-α, GM-CSF, and IL-4. The second pathway or alternative pathway is required for lymphocyte development and is stimulated by a subset of TNF receptors. The third pathway does not require any IKK activation.

IKK- Iκβ kinase.

Iκβ- Inhibitor of NF-κβ.

NEMO- NF-κB essential modulator.

TRAF- Tumor necrosis factor receptor-associated factor.

Figure: NF-κβ signaling pathway:

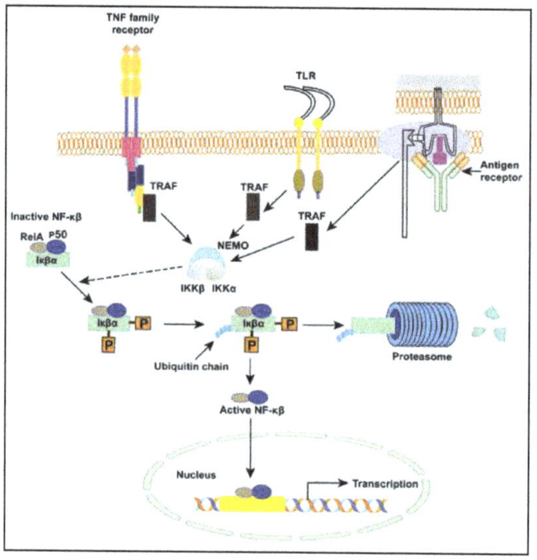

The NF-AT Pathway

The response generated to a T- cell is transmitted from the T cell receptor to a signal transducing complex called CD3 which is followed by clustering of CD3 chains in lipid rafts. Immunoreceptor tyrosine-based, activation motifs (ITAMs) are the specific amino acid sequences in the cytoplasmic domains of CD3 proteins. ITAMs later on become binding sites for several tyrosine kinases on activation. Zeta associated protein – 70 (ZAP-70) is one of the TKs and on phosphorylation can stimulate three signaling pathways. In the first pathway activation of calcineurin, a phosphatase is required and it ends up by generating NF-κβ and NF-AT . In the second pathway, ZAP-70 stimulates a protein kinase C which is required for activation of NF-κβ . The third pathway activates ras, a GTP-binding protein, fos and jun.

The JAK-STAT Pathway

More than 40 cytokines use the JAK-STAT pathway. The stimulated JAK (Janus kinase) molecules phosphorylate tyrosine residues on one of several STAT proteins (signal transducers and activators of transcription). Phosphorylated STAT proteins then dimerize and dissociate from JAK and get shifted to the nucleus where they regulate the expression of target genes.

Figure: JAK-STAT signaling pathway:

Cytokine ligation to cytokine receptor leads to activation of JAK tyrosine kinase, phosphorylation of its cytoplasmic domain, and recruitment of STAT. Recruitment of STAT and its phosphorylation dimerizes and transport it to nucleus. This event activates the transcription of cytokine gene.

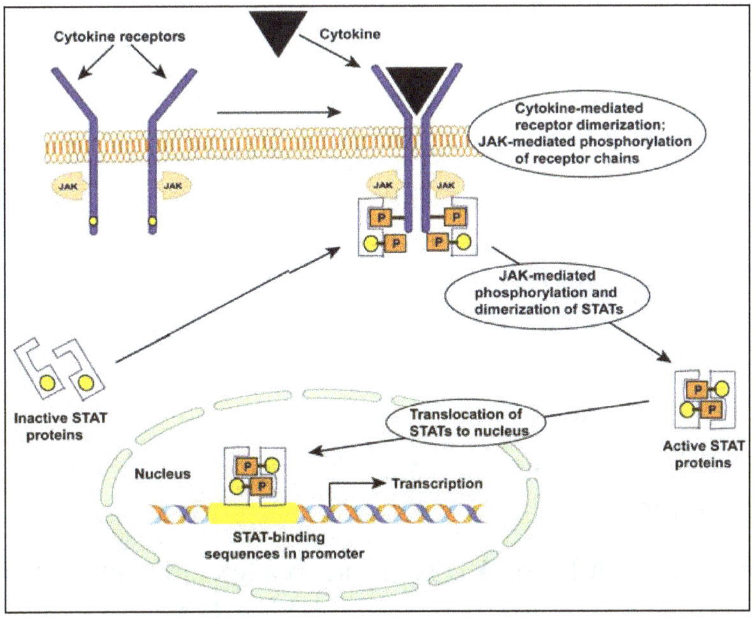

Interferon alpha/beta domain.

Interferon

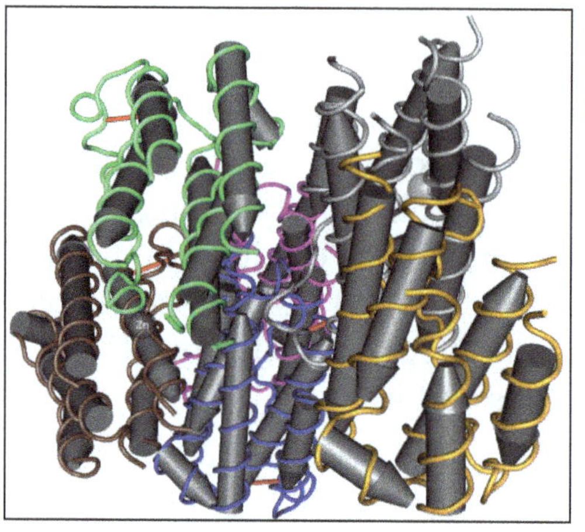

The molecular structure of human interferon-alpha.

Interferons (IFNs) are a group of signaling proteins made and released by host cells in response to the presence of several pathogens, such as viruses, bacteria, parasites, and also tumor cells. In a typical scenario, a virus-infected cell will release interferons causing nearby cells to heighten their anti-viral defenses.

IFNs belong to the large class of proteins known as cytokines, molecules used for communication between cells to trigger the protective defenses of the immune system that help eradicate pathogens. Interferons are named for their ability to "interfere" with viral replication by protecting cells from virus infections. IFNs also have various other functions: they activate immune cells, such as natural killer cells and macrophages; they increase host defenses by up-regulating antigen presentation by virtue of increasing the expression of major histocompatibility complex (MHC) antigens. Certain symptoms of infections, such as fever, muscle pain and "flu-like symptoms", are also caused by the production of IFNs and other cytokines.

More than twenty distinct IFN genes and proteins have been identified in animals, including humans. They are typically divided among three classes: Type I IFN, Type II IFN, and Type III IFN. IFNs belonging to all three classes are important for fighting viral infections and for the regulation of the immune system.

Types of Interferon

Based on the type of receptor through which they signal, human interferons have been classified into three major types.

- Interferon type I: All type I IFNs bind to a specific cell surface receptor complex known as the IFN-α/β receptor (IFNAR) that consists of IFNAR1 and IFNAR2 chains. The type I interferons present in humans are IFN-α, IFN-β,

IFN-ε, IFN-κ and IFN-ω. In general, type I interferons are produced when the body recognizes a virus has invaded it. They are produced by fibroblasts and monocytes. However, the production of type I IFN-α is prohibited by another cytokine known as Interleukin-10. Once released, type I interferons will activate molecules which prevent the virus from producing and replicating its RNA and DNA. Overall, IFN-α can be used to treat hepatitis B and C infections, while IFN-β can be used to treat multiple sclerosis.

- Interferon type II (IFN-γ in humans): This is also known as immune interferon and is activated by Interleukin-12. Furthermore, type II interferons are released by T helper cells, type 1 specifically. However, they block the proliferation of T helper cells type two. The previous results in an inhibition of Th2 immune response and a further induction of Th1 immune response, which leads to the development of debilitating diseases such as multiple sclerosis. IFN type II binds to IFNGR, which consists of IFNGR1 and IFNGR2 chains and has a different receptor than type I IFN.
- Interferon type III: Signal through a receptor complex consisting of IL10R2 (also called CRF2-4) and IFNLR1 (also called CRF2-12). Although discovered more recently than type I and type II IFNs, recent information demonstrates the importance of Type III IFNs in some types of virus infections.

In general, type I and II interferons are responsible for regulating and activating the immune response. Expression of type I and III IFNs can be induced in virtually all cell types upon recognition of viral components, especially nucleic acids, by cytoplasmic and endosomal receptors, whereas type II interferon is induced by cytokines such as IL-12, and its expression is restricted to immune cells such as T cells and NK cells.

Function

All interferons share several common effects: they are antiviral agents and they modulate functions of the immune system. Administration of Type I IFN has been shown experimentally to inhibit tumor growth in animals, but the beneficial action in human tumors has not been widely documented. A virus-infected cell releases viral particles that can infect nearby cells. However, the infected cell can prepare neighboring cells against a potential infection by the virus by releasing interferons. In response to interferon, cells produce large amounts of an enzyme known as protein kinase R (PKR). This enzyme phosphorylates a protein known as eIF-2 in response to new viral infections; the phosphorylated eIF-2 forms an inactive complex with another protein, called eIF2B, to reduce protein synthesis within the cell. Another cellular enzyme, RNAse L—also induced by interferon action—destroys RNA within the cells to further reduce protein synthesis of both viral and host genes. Inhibited protein synthesis destroys both the virus and infected host cells. In addition, interferons induce production of hundreds of other proteins—known collectively as interferon-stimulated genes (ISGs)—that have roles in combating viruses and other actions produced by interferon. They also limit viral spread by increasing p53 activity, which kills virus-infected cells by promoting apoptosis. The effect of IFN on p53 is also linked to its protective role against certain cancers.

Another function of interferons is to upregulate major histocompatibility complex molecules, MHC I and MHC II, and increase immunoproteasome activity. Higher MHC I expression increases presentation of viral peptides to cytotoxic T cells, while the immunoproteasome processes viral peptides for loading onto the MHC I molecule, thereby increasing the recognition and killing of infected cells. Higher MHC II expression increases presentation of viral peptides to helper T cells; these cells release cytokines (such as more interferons and interleukins, among others) that signal to and co-ordinate the activity of other immune cells.

Interferons, such as interferon gamma, directly activate other immune cells, such as macrophages and natural killer cells.

Induction of Interferons

Production of interferons occurs mainly in response to microbes, such as viruses and bacteria, and their products. Binding of molecules uniquely found in microbes—viral glycoproteins, viral RNA, bacterial endotoxin (lipopolysaccharide), bacterial flagella, CpG motifs—by pattern recognition receptors, such as membrane bound Toll like receptors or the cytoplasmic receptors RIG-I or MDA5, can trigger release of IFNs. Toll Like Receptor 3 (TLR3) is important for inducing interferons in response to the presence of double-stranded RNA viruses; the ligand for this receptor is double-stranded RNA (dsRNA). After binding dsRNA, this receptor activates the transcription factors IRF3 and NF-κB, which are important for initiating synthesis of many inflammatory proteins. RNA interference technology tools such as siRNA or vector-based reagents can either silence or stimulate interferon pathways. Release of IFN from cells (specifically IFN-γ in lymphoid cells) is also induced by mitogens. Other cytokines, such as interleukin 1, interleukin 2, interleukin-12, tumor necrosis factor and colony-stimulating factor, can also enhance interferon production.

Downstream Signaling

By interacting with their specific receptors, IFNs activate *signal transducer and activator of transcription* (STAT) complexes; STATs are a family of transcription factors that regulate the expression of certain immune system genes. Some STATs are activated by both type I and type II IFNs. However each IFN type can also activate unique STATs.

STAT activation initiates the most well-defined cell signaling pathway for all IFNs, the classical Janus kinase-STAT (JAK-STAT) signaling pathway. In this pathway, JAKs associate with IFN receptors and, following receptor engagement with IFN, phosphorylate both STAT1 and STAT2. As a result, an IFN-stimulated gene factor 3 (ISGF3) complex forms—this contains STAT1, STAT2 and a third transcription factor called IRF9—and moves into the cell nucleus. Inside the nucleus, the ISGF3 complex binds to specific nucleotide sequences called *IFN-stimulated response elements* (ISREs) in the promoters of certain genes, known as IFN stimulated genes ISGs. Binding of ISGF3 and other transcriptional complexes activated by IFN signaling to these specific regulatory elements induces transcription of those genes. A collection of known ISGs is available on Interferome, a curated online database of ISGs; Additionally, STAT homodimers or heterodimers form from different combinations of STAT-1, -3, -4, -5, or -6 during IFN signaling; these dimers initiate gene transcription by binding to IFN-activated site (GAS) elements in gene promoters. Type I IFNs can induce expression of genes with either ISRE or GAS elements, but gene induction by type II IFN can occur only in the presence of a GAS element.

In addition to the JAK-STAT pathway, IFNs can activate several other signaling cascades. For instance, both type I and type II IFNs activate a member of the CRK family of adaptor proteins called CRKL, a nuclear adaptor for STAT5 that also regulates signaling through the C3G/Rap1 pathway. Type I IFNs further activate *p38 mitogen-activated protein kinase* (MAP kinase) to induce gene transcription. Antiviral and antiproliferative effects specific to type I IFNs result from p38 MAP kinase signaling. The *phosphatidylinositol 3-kinase* (PI3K) signaling pathway is also regulated by both type I and type II IFNs. PI3K activates P70-S6 Kinase 1, an enzyme that increases protein synthesis and cell proliferation; phosphorylates of ribosomal protein s6, which is involved in protein synthesis; and phosphorylates a translational repressor protein called *eukaryotic translation-initiation factor 4E-binding protein 1* (EIF4EBP1) in order to deactivate it.

Virus Resistance to Interferons

Many viruses have evolved mechanisms to resist interferon activity. They circumvent the IFN response by blocking downstream signaling events that occur after the cytokine binds to its receptor, by preventing further IFN production, and by inhibiting the functions of proteins that are induced by IFN. Viruses that inhibit IFN signaling include Japanese Encephalitis Virus (JEV), dengue type 2 virus (DEN-2) and viruses of the herpesvirus family, such as human cytomegalovirus (HCMV) and Kaposi's sarcoma-associated herpesvirus (KSHV or HHV8). Viral proteins proven to affect IFN signaling include EBV nuclear antigen 1 (EBNA1) and EBV nuclear antigen 2 (EBNA-2) from Epstein-Barr virus, the large T antigen of Polyomavirus, the E7 protein of Human papillomavirus (HPV), and the B18R protein of vaccinia virus. Reducing IFN-α activity may prevent signaling via STAT1, STAT2, or IRF9 (as with JEV infection) or through the JAK-STAT pathway (as with DEN-2 infection). Several poxviruses encode soluble IFN receptor homologs—like the B18R protein of the vaccinia virus—that bind to and prevent IFN interacting with its cellular receptor, impeding communication between this cytokine and its target cells. Some viruses can encode proteins that bind to double-stranded RNA (dsRNA) to prevent the activity of RNA-dependent protein kinases; this is the mechanism reovirus adopts using its sigma 3 (σ3) protein, and vaccinia virus employs using the gene product of its E3L gene, p25. The ability of interferon to induce protein production from interferon stimulated genes (ISGs) can also be affected. Production of protein kinase R, for example, can be disrupted in cells infected with JEV Some viruses escape the anti-viral activities of interferons by gene (and thus protein) mutation. The H5N1 influenza virus, also known as bird flu, has resistance to interferon and other anti-viral cytokines that is attributed to a single amino acid change in its Non-Structural Protein 1 (NS1), although the precise mechanism of how this confers immunity is unclear.

Interferon Therapy

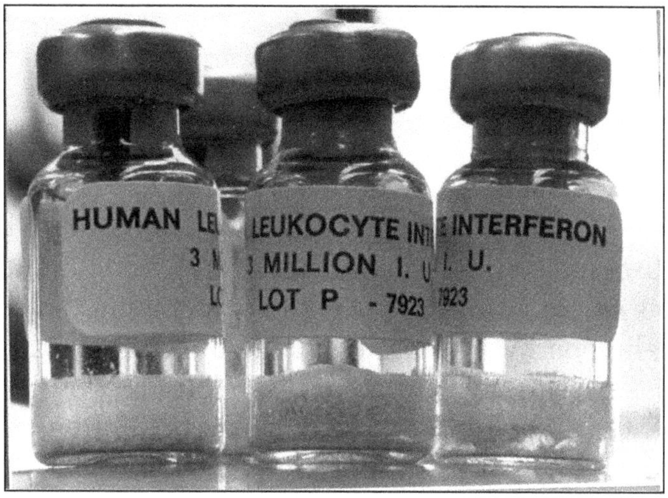

Three vials filled with human leukocyte interferon.

Diseases

Interferon beta-1a and interferon beta-1b are used to treat and control multiple sclerosis, an autoimmune disorder. This treatment is effective for reducing attacks in relapsing-remitting multiple sclerosis and slowing disease progression and activity in secondary progressive multiple sclerosis.

Interferon therapy is used (in combination with chemotherapy and radiation) as a treatment for some cancers. This treatment can be used in hematological malignancy; leukemia and lymphomas including hairy cell leukemia, chronic myeloid leukemia, nodular lymphoma, and cutaneous T-cell lymphoma. Patients with recurrent melanomas receive recombinant IFN-α2b. Both hepatitis B and hepatitis C are treated with IFN-α, often in combination with other antiviral drugs. Some of those treated with interferon have a sustained virological response and can eliminate hepatitis virus. The most harmful strain—hepatitis C genotype I virus—can be treated with a 60-80% success rate with the current standard-of-care treatment of interferon-α, ribavirin and recently approved protease inhibitors such as Telaprevir (Incivek) May 2011, Boceprevir (Victrelis) May 2011 or the nucleotide analog polymerase inhibitor Sofosbuvir (Sovaldi)

December 2013. Biopsies of patients given the treatment show reductions in liver damage and cirrhosis. Some evidence shows giving interferon immediately following infection can prevent chronic hepatitis C, although diagnosis early in infection is difficult since physical symptoms are sparse in early hepatitis C infection. Control of chronic hepatitis C by IFN is associated with reduced hepatocellular carcinoma.

Interferon treatment was evaluated in individuals suffering from herpes simplex virus epithelial keratitis. Topical interferon therapy was shown to be an effective treatment, especially with higher concentrations. Interferon, either used alone or in combination with debridement, appears to be as effective as a nucleoside antiviral agent. The combination of interferon and another nucleoside antiviral agent may speed the healing process.

When used in the systemic therapy, IFNs are mostly administered by an intramuscular injection. The injection of IFNs in the muscle or under skin is generally well tolerated. The most frequent adverse effects are flu-like symptoms: increased body temperature, feeling ill, fatigue, headache, muscle pain, convulsion, dizziness, hair thinning, and depression. Erythema, pain and hardness on the spot of injection are also frequently observed. IFN therapy causes immunosuppression, in particular through neutropenia and can result in some infections manifesting in unusual ways.

Drug formulations

Pharmaceutical forms of interferons	
Generic name	**Trade name**
Interferon alpha 2a	Roferon A
Interferon alpha 2b	Intron A/Reliferon/Uniferon
Human leukocyte Interferon-alpha (HuIFN-alpha-Le)	Multiferon
Interferon beta 1a, liquid form	Rebif
Interferon beta 1a, lyophilized	Avonex
Interferon beta 1a, biogeneric (Iran)	Cinnovex
Interferon beta 1b	Betaseron / Betaferon
Interferon gamma 1b	Actimmune
PEGylated interferon alpha 2a	Pegasys
PEGylated interferon alpha 2a (Egypt)	Reiferon Retard
PEGylated interferon alpha 2b	PegIntron
PEGylated interferon alpha 2b plus ribavirin (Canada)	Pegetron

Several different types of interferons are approved for use in humans. One was first approved for medical use in 1986. For example, in January 2001, the Food and Drug Administration (FDA) approved the use of PEGylated interferon-alpha in the USA; in this formulation, PEGylated interferon-alpha-2b (*Pegintron*), polyethylene glycol is linked to the interferon molecule to make the interferon last longer in the body. Approval for PEGylated interferon-alpha-2a (*Pegasys*) followed in October 2002. These PEGylated drugs are injected once weekly, rather than administering two or three times per week, as is necessary for conventional interferon-alpha. When used with the antiviral drug ribavirin, PEGylated interferon is effective in treatment of hepatitis C; at least 75% of people with hepatitis C genotypes 2 or 3 benefit from interferon treatment, although this is effective in less than 50% of people infected with genotype 1 (the more common form of hepatitis C virus in both the U.S. and Western Europe). Interferon-containing regimens may also include protease inhibitors such as boceprevir and telaprevir.

History

Sidney Pestka of Rutgers University, seen here receiving the National Medal of Technology.

Interferons were first described in 1957 by Alick Isaacs and Jean Lindenmann at the National Institute for Medical Research in London; the discovery was a result of their studies of viral interference. Viral interference refers to the inhibition of virus growth caused by previous exposure of cells to an active or a heat-inactivated virus. Isaacs and Lindenmann were working with a system that involved the inhibition of the growth of live influenza virus in chicken embryo chorioallantoic membranes by heat-inactivated influenza virus. Their experiments revealed that this interference was mediated by a protein released by cells in the heat-inactivated influenza virus-treated membranes. They published their results in 1957 naming the antiviral factor they had discovered *interferon*. The findings of Isaacs and Lindenmann have been widely confirmed and corroborated in the literature.

Furthermore, others may have made observations on interferons before the publication of Isaacs and Lindenmann. For example, during research to produce a more efficient vaccine for smallpox, Yasu-ichi Nagano and Yasuhiko Kojima—two Japanese virologists working at the Institute for Infectious Diseases at the University of Tokyo—noticed inhibition of viral growth in an area of rabbit-skin or testis previously inoculated with UV-inactivated virus. They hypothesised that some "viral inhibitory factor" was present in the tissues infected with virus and attempted to isolate and characterize this factor from tissue homogenates. Independently, Monto Ho, in John Enders's lab, observed in 1957 that attenuated poliovirus conferred a species specific anti-viral effect in human amniotic cell cultures. They described these observations in a publication, naming the responsible factor *viral inhibitory factor* (VIF). It took another fifteen to twenty years, using somatic cell genetics, to show that the interferon action gene and interferon gene reside in different human chromosomes. The purification of human beta interferon did not occur until 1977. Chris Y.H. Tan and his co-workers purified and produced biologically active, radio-labeled human beta interferon by superinducing the interferon gene in fibroblast cells, and they showed its active site contains tyrosine residues. Tan's laboratory isolated sufficient amounts of human beta interferon to perform the first amino acid, sugar composition and N-terminal analyses. They showed that human beta interferon was an unusually hydrophobic glycoprotein. This explained the large loss of interferon activity when preparations were transferred from test tube to test tube or from vessel to vessel during purification. The analyses showed the reality of interferon activity by chemical verification. The purification of human alpha interferon was not reported until 1978. A series of publications from the laboratories of Sidney Pestka and Alan Waldman between 1978 and 1981, describe the purification of the type I interferons IFN-α and IFN-β. By the early 1980s, genes for these interferons had been cloned, adding further definitive proof that interferons were responsible for interfering with viral replication. Gene cloning also confirmed that IFN-α was encoded by a family of many related genes. The type II IFN (IFN-γ) gene was also isolated around this time.

Interferon was scarce and expensive until 1980, when the interferon gene was inserted into bacteria using recombinant DNA technology, allowing mass cultivation and purification from bacterial cultures or derived from yeasts. Interferon can also be produced by recombinant mammalian cells. Before the early 1970s, large scale production of human interferon had been pioneered by Kari Cantell. He produced large amounts of human alpha interferon from large quantities of human white blood cells collected by the Finnish Blood Bank. Large amounts of human beta interferon were made by superinducing the beta interferon gene in human fibroblast cells.

Cantell's and Tan's methods of making large amounts of natural interferon were critical for chemical characterisation, clinical trials and the preparation of small amounts of interferon messenger RNA to clone the human alpha and beta interferon genes. The superinduced human beta interferon messenger RNA was prepared by Tan's lab for Cetus corp. to clone the human beta interferon gene in bacteria and the recombinant interferon was developed as 'betaseron' and approved for the treatment of MS. Superinduction of the human beta interferon gene was also used by Israeli scientists to manufacture human beta interferon.

Human Interferons

- IFNA1.
- IFNA2.
- IFNA4.
- IFNA5.
- IFNA6.
- IFNA7.
- IFNA8.
- IFNA10.
- IFNA13.
- IFNA14.
- IFNA16.
- IFNA17.
- IFNA21.
- IFNB1.
- IFNW.
- IFNE1.
- IFNK.

References

- Royal Swedish Academy of Sciences (10 October 2012). "The Nobel Prize in Chemistry 2012 Robert J. Lefkowitz, Brian K. Kobilka". Retrieved 05, January 2020
- Long, Sarah S.; Pickering, Larry K.; Prober, Charles G. (2012). Principles and Practice of Pediatric Infectious Disease. Elsevier Health Sciences. p. 1502. ISBN 1437727026
- Aggarwal S, Gurney AL (Jan 2002). "IL-17: prototype member of an emerging cytokine family". Journal of Leukocyte Biology. 71 (1): 1–8. PMID 11781375
- Julius M. Cruse; Robert Edwin Lewis (20 April 2009). Illustrated dictionary of immunology. CRC Press. pp. 208–. ISBN 978-0-8493-7987-1. Retrieved 11, February 2020
- Abbas AK, Lichtman AH, Pillai S (2012). Cellular and molecular immunology (7th ed.). Philadelphia: Elsevier/Saunders. ISBN 1437715281
- Kolata, Gina (2015-01-22). "Jean Lindenmann, Who Made Interferon His Life's Work, Is Dead at 90". New York Times. Retrieved 21, March 2020

- "The Top Prescription Drugs of 2012 Globally: Biologics Dominate, But Small Molecule CNS Drugs Hold on to Top Spots" (PDF). ACS Chemical Neuroscience. Retrieved 09, June 2020

- Mélik-Parsadaniantz, Stéphane; Rostène, William (2008). "Chemokines and neuromodulation". Journal of Neuroimmunology. 198 (1–2): 62–8. doi:10.1016/j.jneuroim.2008.04.022. PMID 18538863

Chapter 6

Immune System Disorders

Like any other functional system of the body, the immune system is prone to disorders which affect its functions and can sometimes be fatal. Some disorders which may plague the immune system are autoimmune diseases, hypersensitivity, immunodeficiency, transplant rejection, autoimmunity, HIV/AIDS, etc. This chapter aims to provide an in-depth understanding of these immune system disorders.

Autoimmune Disease

An autoimmune disease is a pathological state arising from an abnormal immune response of the body to substances and tissues that are normally present in the body.

Autoimmunity, on the other hand, is the presence of self-reactive immune response (e.g., auto-antibodies, self-reactive T-cells), with or without damage or pathology resulting from it. This may be restricted to certain organs (e.g. in autoimmune thyroiditis) or involve a particular tissue in different places (e.g. Goodpasture's disease which may affect the basement membrane in both the lung and the kidney).

The treatment of autoimmune diseases is typically with immunosuppression—medication that decreases the immune response. New novel treatments include Cytokine blockade (or the blockade of cytokine signaling pathways), removal of effector T-cells and B-cells (e.g. anti-CD20 therapy can be effective at removing instigating B-cells). Intravenous Immunoglobulin has been helpful in treating some antibody mediated autoimmune diseases as well, possibly through negative feedback mechanisms.

A large number of autoimmune diseases are recognized. A major understanding of the underlying pathophysiology of autoimmune diseases has been the application of genome wide association scans that have identified a striking degree of genetic sharing among the autoimmune diseases.

Definition

For a disease to be regarded as an autoimmune disease it needs to answer to *Witebsky's postulates* (first formulated by Ernest Perez and colleagues in 1957 and modified in 1994):

- Direct evidence from transfer of disease-causing antibody or disease-causing T lymphocyte white blood cells.
- Indirect evidence based on reproduction of the autoimmune disease in experimental animals.
- Circumstantial evidence from clinical clues.
- Genetic evidence suggesting "clustering" with other autoimmune diseases.
- Autoimmune diseases are incurable.

Effects

Autoimmune diseases have a wide variety of different effects. They do tend to have one of three characteristic pathological effects which characterize them as autoimmune diseases:

- Damage to or destruction of tissues.
- Altered organ growth.
- Altered organ function.

It has been estimated that autoimmune diseases are among the leading causes of death among women in the United States in all age groups up to 65 years.

A substantial minority of the population suffers from these diseases, which are often chronic, debilitating, and life-threatening.

There are more than 80 illnesses caused by autoimmunity. Autoimmune diseases affect approximately 2-5% of the western world's population. Women are found to be more commonly affected than men. Environmental events can trigger some cases of autoimmune diseases such as exposure to radiation or certain drugs which can damage tissues of the body. Infection can also be a trigger of some autoimmune diseases for example Lupus which is thought to be a milder version of an idiopathic disorder where there is an increased production of antihistone antibodies.

Pathophysiology

The human immune system typically produces both T-cells and B-cells that are capable of being reactive with self-antigens, but these self-reactive cells are usually either killed prior to becoming active within the immune system, placed into a state of anergy (silently removed from their role within the immune system due to over-activation), or removed from their role within the immune system by regulatory cells. When any one of these mechanisms fail, it is possible to have a reservoir of self-reactive cells that become functional within the immune system. The mechanisms of preventing self-reactive T-cells from being created takes place through Negative selection process within the thymus as the T-cell is developing into a mature immune cell.

Some infections, such as Campylobacter jejuni, have antigens that are similar (but not identical) to our own self-molecules. In this case, a normal immune response to C. jejuni can result in the production of antibodies that also react to a lesser degree with receptors on skeletal muscle (i.e., Myasthenia gravis).

There are many theories as to how an autoimmune disease state arises. Some common ones are listed below.

Cryptic Determinants/Molecular Sequestration

Although it is possible for a potential auto antigen to be geographically sequestered in an immune privileged site within the body (e.g. the eye), mechanisms exist to express even these antigens in a tolerogenic fashion to the immune system. However, it is impossible to induce tolerance (immune unresponsiveness) to all aspects of an autoantigen. This is because under normal physiologic conditions some regions of a self-antigen are not expressed at a sufficient level to induce tolerance. These poorly displayed areas of an antigen are called "cryptic determinants." The immune system maintains a high-affinity repertoire to the cryptic self because the presentation of these determinants was insufficient to induce strong tolerance.

Molecular Mimicry

The concept of molecular mimicry describes a situation in which a foreign antigen can initiate an immune response in which a T or B cell component cross-recognizes self. The cross reactive immune response is responsible for the autoimmune disease state. Cross-reactive immune responses to self were first described for antibodies.

Altered Glycan Theory

According to this theory the effector function of the immune response is mediated by the glycans (polysaccharides) displayed by the cells and humoral components of the immune system. Individuals with autoimmunity have alterations in their glycosylation profile such that a proinflammatory immune response is favored. It is further hypothesized that individual autoimmune diseases will have unique glycan signatures.

Prevalence

The first estimate of US prevalence for autoimmune diseases as a group was published in 1997 by Jacobson, et al. They reported US prevalence to be around 9 million, applying prevalence estimates for 24 diseases to a US population of 279 million. Jacobson's work was updated by Hayter & Cook in 2012. This study used Witebsky's postulates, as revised by Rose & Bona, to extend the list to 81 diseases and estimated overall cumulative US prevalence for the 81 autoimmune diseases at 5.0%, with 3.0% for males and 7.1% for females. The estimated community prevalence, which takes into account the observation that many people have more than one autoimmune disease, was 4.5% overall, with 2.7% for males and 6.4% for females. Based on a US population of 320 million, the Hayter & Cook data indicate US prevalence at about 14.5 million.

In a report to the US Congress, the National Institutes of Health reported prevalence between 14.7 and 23.5 million. Although the NIH report predates the Hayter & Cook study by 7 years, the NIH low-range number closely matches their estimate. However, there is no substantiation provided for the higher estimate, and no list of diseases, or any reference to a list of diseases, is included in that report.

List by Category

This list of autoimmune diseases is categorized by organ and tissue type to help locate diseases that may be similar.

- Major Organs:
  - Heart.
  - Kidney.

- Liver.
- Lung.
- Skin.
- Glands:
    - Endocrine.
    - Adrenal Gland.
    - Multi-glandular.
    - Pancreas.
    - Thyroid Gland.
    - Exocrine.
    - Reproductive Organs.
    - Salivary Glands.
- Digestive System.
- Tissue:
    - Blood.
    - Connective Tissue, Systemic, and multi-organ.
    - Muscle.
    - Nervous System.
    - Eyes.
    - Ears.
    - Vascular system.
- Autoimmune Comorbidities:
- Not Autoimmune:

Other Qualifiers	
A	"Accepted" in prior version of this table
C	A disease, regarded as autoimmune, that is often found in individuals with another autoimmune condition. This designation is given to diseases that are classified by Rose Bona as having "circumstantial" evidence of autoimmune etiology. Diseases in this list with a "C" are, therefore, actual autoimmune diseases, rather than comorbid symptoms, which appear after this list.
E	Disease is an autoimmune response triggered by a specific environmental factor
F	Disease is only caused by autoimmunity in only a fraction of those who suffer from it

I	Described as an autoinflammatory disease
L	Evidence to indicate autoimmunity is extremely limited or circumstantial
M	Disease appears under Autoimmune Diseases in MeSH
N	Not listed in prior version of this table
R	Disease appeared in prior version but has been renamed. In renaming, precedence has been given to scientific names over those based on discoverers.
S	"Suspected" in the prior version of this table
T	Disease has a known trigger, such as viral infection, vaccination, or injury
X	An extremely rare disease, which would suggest limited opportunity to study it and conclusively determine whether it is caused by autoimmunity
Y	Listed in the prior version of this table with "Accepted/Suspected" left blank

Autoimmune Diseases

This is a dynamic list and may never be able to satisfy particular standards for completeness. You can help by expanding it with reliably sourced entries.

Organ/Tissue Type Disease Name	Level of Acceptance for Autoimmunity	Hypersensitivity (I,II,III,IV)	ICD-9 Codes	Notes/Autoantibodies/ Synonyms/Rare Variants
Major Organs<Top>				
Heart<Top>				
Myocarditis	Moderate, F, R, A		391.2422429.0	Synonyms: Autoimmune myocarditis, Autoimmune cardiomyopathy, Coxsackie myocarditis
Postmyocardial infarction syndrome	Limited, R, Y		411.0	Autoantibodies: Myocardial neo-antigens formed as a result of the MI. Synonym: Dressler's syndrome
Postpericardiotomy syndrome	Limited, N		429.4	
Subacute bacterial endocarditis (SBE)	Limited, Y	III	421.0	Autoantibodies: essential mixed cryoglobulinemia.
Kidney<Top>				
Anti-Glomerular Basement Membrane nephritis	Moderate, R, M, A	II	446.21	Autoantibodies: Anti-Basement Membrane Collagen Type IV Protein. Synonyms: Goodpastures Syndrome, Glomerulonephritis Type 1
Interstitial cystitis	Limited, S		595.1	Mast cells.
Lupus nephritis	Comorbidity, N		583.81	A comorbidity of Systemic Lupus Erythematosis.

Liver<Top>				
Autoimmune hepatitis	Moderate, A	cell-mediated	571.42	Autoantibodies: ANA and SMA, LKM-1, LKM-2 or LKM-3; antibodies against soluble liver antigen (anti-SLA, anti-LP) no autoantibodies detected (~20%). Synonym: Lupoid hepatitis
Primary biliary cirrhosis (PBC)	Moderate, A		571.6	Autoantibodies: Anti-p62, Anti-sp100, Anti-Mitochondrial(M2), Anti-Ro aka SSA. Note that Sjogren's is classified in some places (e.g. MeSH) as rheumatoid disease, but there is no published evidence to support that classification.
Primary sclerosing cholangitis	Limited, Y		576.1	Possible overlap with primary biliary cirrhosis. Autoantibodies: HLA-DR52a.
Lung<Top>				
Antisynthetase syndrome	Limited, Y		279.49	
Skin<Top>				
Alopecia Areata	Moderate, A		704.01	Autoantibodies: T-cells. Synonyms: Alopecia areata - Patchy, Totalis, Universalis
Autoimmune Angioedema	Limited, F, N		277.6 995.1	
Autoimmune progesterone dermatitis	Limited, X, A		279.49	
Autoimmune urticaria	Comorbidity, A		708	
Bullous pemphigoid	Moderate, Y		694.5	Autoantibodies: IgG autoantibodies targeting the type XVII collagen component of hemidesmosomes.
Cicatricial pemphigoid	Limited, R, X, Y		694.61	precipitates C3. Autoantibodies: anti-BP-1, anti-BP-2. Synonyms: Benign Mucosal Pemphigoid, Ocular cicatricial pemphigoid
Dermatitis herpetiformis	Moderate, C, Y		694.0	Autoantibodies: IgA Eosinophilia; anti-epidermal transglutaminase antibodies.
Discoid lupus erythematosus	Limited, Y	III	695.4	IL-2 and IFN-gamma.
Epidermolysis bullosa acquisita	Moderate, Y		694.8	COL7A1.
Erythema nodosum	Limited, F, Y		695.2	

Gestational pemphigoid	Limited, R, Y		646.8	Autoantibodies: IgG and C3 misdirected antibodies intended to protect the placenta.
Hidradenitis suppurativa	Limited, C, S		705.83	
Lichen planus	Limited, Y		697.0	
Lichen sclerosus	Limited, C, Y		701.0	
Linear IgA disease (LAD)	Moderate, Y		646.8	
Morphea	Limited, C, S		701.0	
Pemphigus vulgaris	Moderate, M, A	II	694.4	Autoantibodies: Anti-Desmoglein 3 eosinophilia.
Pityriasis lichenoides et varioliformis acuta	Limited, C		696.2	
Mucha-Habermann disease	Limited, C, Y		696.2	T-cells. Synonyms: Pityriasis lichenoides, varioliformis acuta
Psoriasis	Moderate, A	IV?	696	CD-8 T-cells, HLA-Cw6, IL-12b, IL-23b, TNFalpha, NF-κB.
Systemic scleroderma	Limited, R, S		710.1	COL1A2 and TGF-β1. Autoantibodies: Anti-nuclear antibodies, anti-centromere and anti-scl70/anti-topoisomerase antibodies. Synonyms: Diffuse cutaneous systemic sclerosis, Systemic sclerosis, Scleroderma
Vitiligo	Limited, C, S		709.01	NALP-1 RERE, PTPN22, LPP, IL2RA, GZMB, UBASH3A and C1QTNF6.
Glands<Top>				
Endocrine<Top>				
Adrenal Gland<Top>				
Addison's disease	Moderate, F, Y		255	Autoantibodies: 21 hydroxylase.
Multi-glandular<Top>				
Autoimmune polyendocrine syndrome (APS) type 1	Moderate, A	Unknown or multiple	258.1	Synonyms: Whitaker's Syndrome, Autoimmune polyendocrinopathy-candidiasis-ectodermal dystrophy (APECED), Addisons Disease, Polyglandular Autoimmune Syndrome 1 (PGAS-1).

Autoimmune polyendocrine syndrome (APS) type 2	Moderate, A		258.1	DQ2, DQ8 and DRB1*0404. Autoantibodies: anti-21 hydroxylase, anti-17 hydroxylase. Synonyms: Schmidt syndrome, Polyglandular Autoimmune Syndrome 2 (PGAS-2).
Autoimmune polyendocrine syndrome (APS) type 3	Moderate, A		258.1	Synonym: Polyglandular Autoimmune Syndrome 3 (PGAS-3).
Pancreas<Top>				
Autoimmune pancreatitis (AIP)	Moderate, A		577.1	Autoantibodies: ANA, anti-lactoferrin antibodies, anti-carbonic anhydrase antibodies, rheumatoid factor.
Diabetes mellitus type 1	Moderate, A	IV	250.01	HLA-DR3, HLA-DR4. Autoantibodies: Glutamic acid decarboxylase antibodies (GADA), islet cell antibodies (ICA), insulinoma-associated autoantibodies (IA-2), anti-insulin antibodies.
Thyroid Gland<Top>				
Autoimmune thyroiditis	Strong, A	IV	245.8	HLADR5, CTLA-4. Autoantibodies: Antibodies against thyroid peroxidase and/or thyroglobulin. Synonyms: Chronic lymphocytic thyroiditis, Hashimoto's thyroiditis.
Ord's thyroiditis	Moderate, Y		245.8	
Graves' disease	Moderate, M, A	II	242.0	Autoantibodies: Thyroid autoantibodies (TSHR-Ab) that activate the TSH-receptor (TSHR).
Exocrine<Top>				
Reproductive Organs<Top>				
Autoimmune Oophoritis	Moderate, N		614.2	
Endometriosis	Limited, S		617.0	
Autoimmune orchitis	Limited, N		604.0	
Salivary Glands<Top>				
Sjogren's syndrome	Moderate, A		710.2	Autoantibodies: Anti-Ro (often present also in Systemic Lupus Erythematosus).
Digestive System<Top>				
Autoimmune enteropathy	Moderate, X, Y			

Coeliac disease	Moderate, A,E	IV??	579.0	HLA-DQ8 and DQ2.5. Autoantibodies: Anti-tissue transglutaminase antibodies, anti-endomysial IgA, anti-gliadin IgA.
Crohn's disease	Moderate, Y	IV	555	Innate immunity; Th17; Th1; ATG16L1; CARD15; XBP1.
Microscopic colitis	Limited, S		558.9	
Ulcerative colitis	Limited, A	IV	556	
Tissue<Top>				
Blood<Top>				
Antiphospholipid syndrome (APS, APLS)	Moderate, M, A		289.81	HLA-DR7, HLA-B8, HLA-DR2, HLA-DR3. Autoantibodies: Anti-cardiolipin; anti-pyruvate dehydrogenase; β2 glycoprotein I; phosphatidylserine; anti-apoH; Annexin A5. Synonym: Hughes syndrome.
Aplastic anemia	Limited, F, Y		284	
Autoimmune hemolytic anemia	Moderate, M, A	II	283.0	Complement activation.
Autoimmune lymphoproliferative syndrome	Moderate, A		279.41	TNFRSF6; defective Fas-CD95 apoptosis. Synonym: Canale-Smith syndrome.
Autoimmune neutropenia	Moderate, F, N		288.09	
Autoimmune thrombocytopenic purpura	Moderate, M, R, A		287.31	Autoantibodies: Anti gpIIb-IIIa or 1b-IX. Synonym: Idiopathic Thrombocytopenic Purpura (ITP)
Cold agglutinin disease	Moderate, M, A	II	283.0	Idiopathic or secondary to leukemia or infection. Autoantibodies: IgM. Synonym: Autoimmune hemolytic anemia.
Essential mixed cryoglobulinemia	Limited, C, Y		273.2	
Evans syndrome	Moderate, Y		287.32	Syndrome with a combination of hemolytic anemia and thrombocytopenic purpura
Paroxysmal nocturnal hemoglobinuria	Limited, F, S		283.2	
Pernicious anemia	Moderate, A	II	281.0	Autoantibodies: Anti-parietal cell antibody.
Pure red cell aplasia	Limited, Y		284.81	

Thrombocytopenia	Limited, F, Y	II	287.5	Multiple mechanisms. Autoantibodies: Glycoproteins IIb-IIIa or Ib-IX in ITP anti-ADAMTS13 in TTP. and HUS anti-cardiolipin (anti-cardiolipin antibodies) and β2 glycoprotein I in Antiphospholipid syndrome; anti-HPA-1a, anti-HPA-5b, and others in NAIT. Synonyms: Neonatal thrombocytopenia
Connective Tissue, Systemic, and multi-organ<Top>				
Adiposis dolorosa	Limited, L, S		272.8	Lipoid tissue. Synonym: Dercum's disease
Adult-onset Still's disease	Moderate, Y		714.2	Macrophage migration inhibitory factor. Autoantibodies: ANA.
Ankylosing Spondylitis	Limited, S		720.0	CD8; HLA-B27.
CREST syndrome	Limited, Y		710.1	Autoantibodies: Anti-centromere antibodies Anti-nuclear antibodies.
Drug-induced lupus	Moderate, Y		710.0	Autoantibodies: Anti-histone antibodies.
Enthesitis-related arthritis	Limited, C, Y			MMP3, TRLR2, TLR4, ERAP1. A subtype of Juvenile Rheumatoid Arthritis.
Eosinophilic fasciitis	Limited, F, A		728.89	Synonym: Shulman's syndrome
Felty syndrome	Strong, M, Y		714.1	
IgG4-related disease	Limited, C, N			Characteristic histological features (storiform fibrosis, lymphoplasmacytic infiltrate, obliterative phlebitis) are required for definitive diagnosis. Synonyms: IgG4-related systemic disease, IgG4-related sclerosing disease, IgG4-related systemic sclerosing disease, IgG4-related autoimmune disease, IgG4-associated multifocal systemic fibrosis, IgG4-associated disease, IgG4 syndrome, Hyper-IgG4 disease, Systemic IgG4-related plasmacytic syndrome.
Juvenile Arthritis	Strong, M, R, Y		714.30	Autoantibodies: inconsistent ANA, Rheumatoid factor. Synonyms: Juvenile rheumatoid arthritis, Juvenile idiopathic arthritis

Lyme disease (Chronic)	Limited, L, T, N		088.81	
Mixed connective tissue disease (MCTD)	Moderate, M, A		710.8	HLA-DR4. Autoantibodies: Anti-nuclear antibody, anti-U1-RNP.
Palindromic rheumatism	Limited, Y		719.3	Autoantibodies: Anti-cyclic citrullinated peptide antibodies (anti-CCP) and antikeratin antibodies (AKA). Synonym: Hench-Rosenberg syndrome.
Parry Romberg syndrome	Limited, Y		349.89351.8	Autoantibodies: ANA.
Parsonage-Turner syndrome	Limited, Y		353.5	
Psoriatic arthritis	Moderate, C, A	IV?	696.0	HLA-B27.
Reactive arthritis	Limited, C, F, Y		099.3	Synonym: Reiter's syndrome
Relapsing polychondritis	Strong, A		733.99	Synonyms: Atrophic polychondritis, systemic chondromalacia, chronic atrophic polychondritis, Meyenburg-Altherr-Uehlinger syndrome, generalized chondromalacia, systemic chondromalacia
Retroperitoneal fibrosis	Limited, Y		593.4	
Rheumatic fever	Moderate, T, A	II	390	Autoantibodies: Streptococcal M protein cross reacts with human myosin.
Rheumatoid arthritis	Strong, M, A	III	714	HLA-DR4, PTPN22, depleted B cells, TNF alpha, IL-17, (also maybe IL-1, 6, and 15). Autoantibodies: Rheumatoid factor (anti-IgGFc), Anti-MCV, ACPAs(Vimentin).
Sarcoidosis	Limited, S	IV	135	BTNL2; HLA-B7-DR15; HLA DR3-DQ2.
Schnitzler syndrome	Limited, L, X, Y		273.1	IgM.
Systemic Lupus Erythematosus (SLE)	Strong, M, A	III	695.4	Autoantibodies: Anti-nuclear antibodies, anti-Ro (often present also in Sjogren's syndrome). Eosinophilia. Synonym: Lupus
Undifferentiated connective tissue disease (UCTD)	Moderate, C, A		710.9	HLA-DR4. Autoantibodies: anti-nuclear antibody. Synonyms: Latent lupus, incomplete lupus

Muscle<Top>				
Dermatomyositis	Moderate, F, X, A		710.3	B- and T-cell perivascular inflammatory infiltrate on muscle biopsy. Autoantibodies: histidine-tRNA anti-signal recognition peptide Anti-Mi-2 Anti-Jo1. Synonym: Juvenile dermatomyositis
Fibromyalgia	Limited, C, F, N		729.1	
Inclusion body myositis	Limited, F, Y		359.71	Similar to polymyositis, but does not respond to steroid therapy-activated T8 cells.
Myositis	Limited, F, Y		729.1	
Myasthenia gravis	Strong, M, A	II	358	HA-B8 HLA-DR3 HLA-DR1. Autoantibodies: Nicotinic acetylcholine receptor MuSK protein.
Neuromyotonia	Limited, F, S	II?	333.90	Autoantibodies: Voltage-gated potassium channels. Synonym: Isaacs' syndrome
Paraneoplastic cerebellar degeneration	Limited, Y	IV? II?	334.9	Autoantibodies: anti-Yo (anti-cdr-2 in purkinje fibers) anti-Hu, anti-Tr, antiglutamate receptor.
Polymyositis	Limited, F, A		710.4	Autoantibodies: IFN-gamma, IL-1, TNF-alpha.
Nervous System<Top>				
Acute disseminated encephalomyelitis (ADEM)	Strong, M, T, A		323.61323.81	Synonyms: Perivenous encephalomyelitis, Acute hemorrhagic leukoencephalitis (AHL, AHLE), Acute necrotizing encephalopathy (ANE), Acute hemorrhagic encephalomyelitis (AHEM), Acute necrotizing hemorrhagic leukoencephalitis (ANHLE), Weston-Hurst syndrome, Hurst's disease.
Acute motor axonal neuropathy	Limited, N		356.8	
Anti-N-Methyl-D-Aspartate (Anti-NMDA) Receptor Encephalitis	Moderate, N			
Balo concentric sclerosis	Moderate, Y		341.1	Synonyms: Balo disease, Schilders disease.
Bickerstaff's encephalitis	Limited, Y		323.62	Similar to Guillain-Barré syndrome. Autoantibodies: Anti-GQ1b 2/3 patients.

Chronic inflammatory demyelinating polyneuropathy	Moderate, C, Y		357.81	Similar to Guillain–Barré syndrome. Autoantibodies: Anti-ganglioside antibodies. Synonyms: Relapsing polyneuropathy (CRP), chronic inflammatory demyelinating polyradiculoneuropathy, Chronic inflammatory demyelinating polyneuritis.
Guillain–Barré syndrome	Strong, M, A	IV	357.0	Autoantibodies: Anti-ganglioside, anti-GQ1b. Synonyms: Miller-Fisher syndrome, Landry's paralysis.
Hashimoto's encephalopathy	Moderate, C, X, A	IV		Autoantibodies: Alpha-enolase. Synonyms: Steroid-responsive encephalopathy associated with autoimmune thyroiditis (SREAT), Nonvasculitic autoimmune meningoencephalitis (NAIM), Encephalopathy Associated with Autoimmune Thyroid Disease (EAATD).
Idiopathic inflammatory demyelinating diseases	Limited, F, Y		356.8	A set of different variants of multiple sclerosis.
Lambert-Eaton myasthenic syndrome	Strong, M, Y		358.1	HLA-DR3-B8. Autoantibodies: Voltage-gated calcium channels; Q-type calcium channel, synaptogagmin, muscarinic acetylcholine receptor M1.
Multiple sclerosis, pattern II	Strong, M, A	IV	340	Autoantibody against potassium channel has been reported to present demyelination pattern II. Other cases present autoimmunity against MOG and Anoctamin-2. The three reported autoimmune variants belong to MS pattern II. Also involved HLA-DR2, PECAM-1, Anti-myelin basic protein. Autoantibodies: Anti-Kir4.1, Anti-MOG, Anti-ANO2 (heterogeneous). Synonyms: Primary progressive multiple sclerosis, Relapsing-remitting multiple sclerosis, disseminated sclerosis, encephalomyelitis disseminata.

Oshtoran Syndrome	X		F06.9	Heritable, abnormalities in the kynurenine and glutamate metabolism.
Pediatric Autoimmune Neuropsychiatric Disorder Associated with Streptococcus (PANDAS)	Limited, F, S	II?	279.49	Antibodies against streptococcal infection serve as auto-antibodies.
Progressive inflammatory neuropathy	Limited, X, S		356.4	Similar to Guillain-Barré syndrome. Autoantibodies: Anti-ganglioside antibodies:anti-GM1, anti-GD1a, anti-GQ1b.
Restless leg syndrome	Limited, C, S		333.94	May occur in Sjogren's syndrome, coeliac disease and rheumatoid arthritis, or in derangements of iron metabolism.
Stiff person syndrome	Limited, S		333.91	GLRA1 (glycine receptor). Autoantibodies: Glutamic acid decarboxylase (GAD).
Sydenham chorea	Limited, T, Y		392	
Transverse myelitis	Limited, M, A		323.82341.2	
		Eyes<Top>		
Autoimmune retinopathy	Limited, X, N			
Autoimmune uveitis	Moderate, F, A		364	Autoantibodies: HLAB-27.
Cogan syndrome	Limited, F, Y		370.52	
Graves ophthalmopathy	Moderate, M, N		242.9	
Intermediate uveitis	Limited, L, Y		364.3	Synonyms: Pars planitis, Peripheral Uveitis.
Ligneous conjunctivitis	Limited, L, N		372.39	
Mooren's ulcer	Limited, L, N		370.07	
Neuromyelitis optica	Limited, M, Y	II?	341.0	Autoantibodies: NMO-IgG aquaporin 4. Synonym: Devic's disease.
Opsoclonus myoclonus syndrome	Limited, X, S	IV?	379.59	Lymphocyte recruitment to CSF.
Optic neuritis	Limited, C, Y		377.30	
Scleritis	Limited, C, Y		379.0	
Susac's syndrome	Limited, C, Y		348.39	Synonym: Retinocochleocerebral Vasculopathy.
Sympathetic ophthalmia	Limited, I, Y		360.11	Autoantibodies: ocular antigens following trauma.
Tolosa-Hunt syndrome	Limited, I, X, Y		378.55	

Ears<Top>				
Autoimmune inner ear disease (AIED)	Limited, A		388.8	
Ménière's disease	Limited, Y	III?	386.00	Autoantibodies: Major peripheral myelin protein P0.
Vascular system<Top>				
Behçet's disease	Limited, I, X, A		136.1	An immune-mediated systemic vasculitis; linkage to HLA-B51 (HLA-B27); very variable manifestations, with ulcers as common symptom. Synonyms: Morbus Adamandiades-Behçet. Rare Variant: Hughes-Stovin syndrome.
Eosinophilic granulomatosis with polyangiitis (EGPA)	Limited, I, X, Y		446.4	Autoantibodies: p-ANCA Eosinophilia. Synonym: Churg-Strauss syndrome.
Giant cell arteritis	Limited, I, R, A	IV	446.5	Synonyms: Cranial arteritis, Temporal Arteritis.
Granulomatosis with polyangiitis (GPA)	Strong, M, A		446.4	Autoantibodies: Anti-neutrophil cytoplasmic (cANCA). Synonym: Wegener's granulomatosis.
IgA vasculitis (IgAV)	Limited, L, Y		287.0	Autoantibodies: IgA and complement component 3 (C3). Synonyms: Anaphylactoid purpura, Henoch-Schonlein purpura, purpura rheumatica, Schönlein–Henoch purpura.
Kawasaki's disease	Moderate, S,E		446.1	ITPKC HLA-B51. Synonyms: Kawasaki syndrome, lymph node syndrome, mucocutaneous lymph node syndrome.
Leukocytoclastic vasculitis	Limited, L, Y		447.6	
Lupus vasculitis	Moderate, C, N		583.81	A comorbidity of Systemic Lupus Erythematosis.
Rheumatoid vasculitis	Moderate, C, N		447.6	A symptom of Lupus.
Microscopic polyangiitis (MPA)	Limited, Y		446.0	Binds to neutrophils causing them to degranulate and damages endothelium. Autoantibodies: p-ANCA myeloperoxidase. Synonyms: Microscopic polyarteritis, microscopic polyarteritis nodosa.

Polyarteritis nodosa (PAN)	Limited, L, Y		446.0	Synonyms: Panarteritis nodosa, periarteritis nodosa, Kussmaul disease, Kussmaul-Maier disease.
Polymyalgia rheumatica	Limited, L, Y		725	
Urticarial vasculitis	Limited, X, Y	II?	708.9	Clinically may resemble type I hypersensitivity. Autoantibodies: anti C1q antibodies.
Vasculitis	Strong, I, M, F, A	III	447.6	Autoantibodies: ANCA (sometimes).
Systemic<Top>				
Primary Immune Deficiency	Limited, N			

Autoimmune Comorbidities

This list includes conditions that are not diseases but signs common to autoimmune disease. Some, such as Chronic Fatigue Syndrome, are controversial. These conditions are included here because they are frequently listed as autoimmune diseases but should not be included in the list above until there is more consistent evidence.

Organ/Tissue Type Disease Name	Level of Acceptance for Autoimmunity	Hypersensitivity (I,II,III,IV)	ICD-9 Codes	Notes/Autoantibodies/Synonyms
Chronic fatigue syndrome	Comorbidity, N			Symptomatic of autoimmune diseases or autoimmune activity, but not a disease or a cause of disease.
Complex regional pain syndrome	Comorbidity, N			Symptomatic of autoimmune diseases or autoimmune activity, but not a disease or a cause of disease. Synonyms: Amplified Musculoskeletal Pain Syndrome, Reflex Neurovascular Dystrophy, Reflex sympathetic dystrophy
Eosinophilic esophagitis	Comorbidity, N		530.13	
Gastritis	Comorbidity, Y			Possibly symptomatic of autoimmune diseases, but not a disease or a cause of disease. Autoantibodies: serum antiparietal and anti-IF antibodies.
Interstitial lung disease	Comorbidity, N			Associated with several autoimmune connective tissue diseases.
POEMS syndrome	Comorbidity, Y			Possibly symptomatic of autoimmune diseases, but not a disease or a cause of disease. Autoantibodies: interleukin 1β, interleukin 6 and TNFα. vascular endothelial growth factor (VEGF), given the ..

Raynaud's phenomenon	Comorbidity, S			Symptomatic of autoimmune diseases or autoimmune activity, but not a disease or a cause of disease.
Primary immunodeficiency	Comorbidity, N		279.8	The condition is inherited, but it is associated with several autoimmune diseases.
Pyoderma gangrenosum	Comorbidity, Y			Possibly symptomatic of autoimmune diseases, but not a disease or a cause of disease.

Not Autoimmune

At this time, there is not sufficient evidence - direct, indirect, or circumstantial - to indicate that these diseases are caused by autoimmunity. These conditions are included here because:

1. The disease was listed in the prior version of this table
2. The disease is included in several widely used lists of autoimmune disease and is shown here to ensure that a person visiting this page does not conclude that the disease was not considered.

Organ/Tissue Type Disease Name	Level of Acceptance for Autoimmunity	Hypersensitivity (I,II,III,IV)	ICD-9 Codes	Notes/Autoantibodies/Synonyms
Agammaglobulinemia	Not Autoimmune, Y		279.00	An immune system disorder but not an autoimmune disease.. Autoantibodies: IGHM; IGLL1: CD79A; CD79B; BLNK; LRRC8A.
Amyloidosis	Not Autoimmune, N		277.30	No consistent evidence of association with autoimmunity.
Amyotrophic lateral sclerosis	Not Autoimmune, Y		335.20	No consistent evidence of association with autoimmunity. Autoantibodies: Amyotrophic lateral sclerosis (Also Lou Gehrig's disease; Motor Neuron Disease).
Anti-tubular basement membrane nephritis	Not Autoimmune, N			No consistent evidence of association with autoimmunity.
Atopic allergy	Not Autoimmune, Y	I	691.8	A hypersensitivity.
Atopic dermatitis	Not Autoimmune, Y	I	691.8	A hypersensitivity.
Autoimmune peripheral neuropathy	Not Autoimmune, F, A			A class of diseases, some of which may be autoimmune. See specific diseases that are listed as autoimmune..
Blau syndrome	Not Autoimmune, Y			Overlaps both sarcoidosis and granuloma annulare. No evidence of association with autoimmunity.

Cancer	Not Autoimmune, Y			No consistent evidence of association with autoimmunity.
Castleman's disease	Not Autoimmune, Y			An immune system disorder but not an autoimmune disease.. Autoantibodies: Over expression of IL-6.
Chagas disease	Not Autoimmune, S			No consistent evidence of association with autoimmunity.
Chronic obstructive pulmonary disease	Not Autoimmune, S			No consistent evidence of association with autoimmunity.
Chronic recurrent multifocal osteomyelitis	Not Autoimmune, Y			LPIN2, D18S60. Synonyms: Majeed syndrome
Complement component 2 deficiency	Not Autoimmune, Y			Possibly symptomatic of autoimmune diseases, but not a disease.
Congenital heart block	Not Autoimmune, N			May be related to autoimmune activity in the mother.
Contact dermatitis	Not Autoimmune, Y	IV		A hypersensitivity.
Cushing's syndrome	Not Autoimmune, Y			No consistent evidence of association with autoimmunity.
Cutaneous leukocytoclastic angiitis	Not Autoimmune, Y			No consistent evidence of association with autoimmunity. Autoantibodies: neutrophils.
Dego's disease	Not Autoimmune, Y			No consistent evidence of association with autoimmunity.
Eczema	Not Autoimmune, Y			No consistent evidence of association with autoimmunity. Autoantibodies: LEKTI, SPINK5, filaggrin., Brain-derived neurotrophic factor (BDNF) and Substance P..
Eosinophilic gastroenteritis	Not Autoimmune, Y			Possibly a hypersensitivity. Autoantibodies: IgE, IL-3, IL-5, GM-CSF, eotaxin.
Eosinophilic pneumonia	Not Autoimmune, F, Y			A class of diseases, some of which may be autoimmune. Specifically, Churg-Strauss syndrome, a subtype of Eosinophilic pneumonia, is autoimmune.
Erythroblastosis fetalis	Not Autoimmune, Y	II		Mother's immune system attacks fetus. An immune system disorder but not autoimmune. Autoantibodies: ABO, Rh, Kell antibodies.
Fibrodysplasia ossificans progressiva	Not Autoimmune, Y			Possibly an immune system disorder but not autoimmune. Autoantibodies: ACVR1 Lymphocytes express increased BMP4.

Gastrointestinal pemphigoid	Not Autoimmune, A			No consistent evidence of association with autoimmunity.
Hypogammaglobulinemia	Not Autoimmune, Y			An immune system disorder but not autoimmune. Autoantibodies: IGHM, IGLL1, CD79A, BLNK, LRRC8A, CD79B.
Idiopathic giant-cell myocarditis	Not Autoimmune, N			No consistent evidence of autoimmune cause though the disease has been found comorbid with other autoimmune diseases. Synonyms: Giant cell myocarditis
Idiopathic pulmonary fibrosis	Not Autoimmune, Y			Autoantibodies: SFTPA1, SFTPA2, TERT, and TERC.. Synonyms: Fibrosing alveolitis
IgA nephropathy	Not Autoimmune, Y	III?		Autoantibodies: IgA produced from marrow rather than MALT. Synonyms: IgA nephrits, Berger's disease, Synpharyngitic Glomerulonephritis. An immune system disorder but not an autoimmune disease.
Immunoregulatory lipoproteins	Not Autoimmune, N			Not a disease.
IPEX syndrome	Not Autoimmune, N			A genetic mutation in FOXP3 that leads to autoimmune diseases, but no consistent evidence that it is an autoimmune disorder itself.. Synonyms: X-linked polyendocrinopathy, immunodeficiency and diarrhea-syndrome (XLAAD)
Ligneous conjunctivitis	Not Autoimmune, N			No consistent evidence of association with autoimmunity.
Majeed syndrome	Not Autoimmune, Y			No consistent evidence of association with autoimmunity. Autoantibodies: LPIN2.
Narcolepsy	Not Autoimmune, Y	II?		No evidence of association with autoimmunity. Research not reproducible. Autoantibodies: hypocretin or orexin, HLA-DQB1*0602.
Rasmussen's encephalitis	Not Autoimmune, Y			No consistent evidence of association with autoimmunity. Autoantibodies: anti-NR2A antibodies.
Schizophrenia	Not Autoimmune, S			No consistent evidence of association with autoimmunity.
Serum sickness	Not Autoimmune, Y	III		A hypersensitivity.
Spondyloarthropathy	Not Autoimmune, Y			No consistent evidence of association with autoimmunity. Autoantibodies: HLA-B27.

Sweet's syndrome	Not Autoimmune, Y			No consistent evidence of association with autoimmunity. Autoantibodies: GCSF.
Takayasu's arteritis	Not Autoimmune, Y			No consistent evidence of association with autoimmunity.
Undifferentiated spondyloarthropathy	Not Autoimmune, Y			See Enthesitis-related arthritis.

Development of Therapies

In both autoimmune and inflammatory diseases, the condition arises through aberrant reactions of the human adaptive or innate immune systems. In autoimmunity, the patient's immune system is activated against the body's own proteins. In chronic inflammatory diseases, neutrophils and other leukocytes are constitutively recruited by cytokines and chemokines, leading to tissue damage.

Mitigation of inflammation by activation of anti-inflammatory genes and the suppression of inflammatory genes in immune cells is a promising therapeutic approach.

Hypersensitivity

Hypersensitivity (also called hypersensitivity reaction or intolerance) is a set of undesirable reactions produced by the normal immune system, including allergies and autoimmunity. These reactions may be damaging, uncomfortable, or occasionally fatal. Hypersensitivity reactions require a pre-sensitized (immune) state of the host. They are classified in four groups after the proposal of P. G. H. Gell and Robin Coombs in 1963.

Coombs and Gell Classification

Comparison of hypersensitivity types				
Type	Alternative names	Often mentioned disorders	Mediators	Description
I	Allergy (immediate)	• Atopy • Anaphylaxis • Asthma	• IgE	Fast response which occurs in minutes, rather than multiple hours or days. Free antigens cross link the IgE on mast cells and basophils which causes a release of vasoactive biomolecules. Testing can be done via skin test for specific IgE.

II	Cytotoxic, antibody-dependent	• Autoimmune hemolytic anemia • Rheumatic heart disease • Thrombocytopenia • Erythroblastosis fetalis • Goodpasture's syndrome • Graves' disease *see type V explanation below • Myasthenia gravis *see type V explanation below	• IgM or IgG • (Complement) • MAC	Antibody (IgM or IgG) binds to antigen on a target cell, which is actually a host cell that is perceived by the immune system as foreign, leading to cellular destruction via the MAC. Testing includes both the direct and indirect Coombs test.
III	Immune complex disease	• Serum sickness • Arthus reaction • Rheumatoid arthritis • Post streptococcal glomerulonephritis • Membranous nephropathy • Lupus nephritis • Systemic lupus erythematosus • Extrinsic allergic alveolitis (hypersensitivity pneumonitis)	• IgG • (Complement) • Neutrophils	Antibody (IgG) binds to soluble antigen, forming a circulating immune complex. This is often deposited in the vessel walls of the joints and kidney, initiating a local inflammatory reaction.
IV	Delayed-type hypersensitivity, cell-mediated immune memory response, antibody-independent	• Contact dermatitis, including Urushiol-induced contact dermatitis (poison ivy rash). • Mantoux test • Chronic transplant rejection • Multiple sclerosis	• T-cells	Helper T cells (specifically Th1 helper t cells) are activated by an antigen presenting cell. When the antigen is presented again in the future, the memory Th1 cells will activate macrophages and cause an inflammatory response. This ultimately can lead to tissue damage.
V	Autoimmune disease, receptor mediated (see below)	• Graves' disease • Myasthenia gravis	• IgM or IgG • (Complement)	

Type V

This is an additional type that is sometimes (often in the UK) used as a distinction from Type 2.

Instead of binding to cell surfaces, the antibodies recognise and bind to the cell surface receptors, which either prevents the intended ligand binding with the receptor or mimics the effects of the ligand, thus impairing cell signaling.

Some clinical examples:

- Graves' disease.
- Myasthenia gravis.

The use of Type 5 is rare. These conditions are more frequently classified as Type 2, though sometimes they are specifically segregated into their own subcategory of Type 2.

Immunodeficiency

Immunodeficiency (or immune deficiency) is a state in which the immune system's ability to fight infectious disease is compromised or entirely absent. Most cases of immunodeficiency are acquired ("secondary") due to extrinsic factors that affect the patient's immune system. Examples of these extrinsic factors include infections, such as by Human Immunodeficiency virus (HIV), extremes of age and environmental factors, such as nutrition. In the clinical setting, the immuno-suppression quality of some drugs, such as steroids, can be utilised. Examples of such use is in transplant surgery as an anti-rejection measure and in patients suffering from an over-active immune system. Some people are born with defects in their immune system, or primary immunodeficiency. A person who has an immunodeficiency of any kind is said to be immunocompromised. An immunocompromised person may be particularly vulnerable to opportunistic infections, in addition to normal infections that could affect everyone. Immunodeficiency may also decrease cancer immunosurveillance

Types

By Affected Component

- Humoral immune deficiency, with signs or symptoms depending on the cause, but generally include signs of hypogammaglobulinemia (decrease of one or more types of antibodies) with presentations including repeated mild respiratory infections, and/or agammaglobulinemia (lack of all or most antibody production) which results in frequent severe infections and is often fatal.
- T cell deficiency, often causes secondary disorders such as acquired immune deficiency syndrome (AIDS).
- *Granulocyte deficiency*, including decreased numbers of granulocytes (called as granulocytopenia or, if absent, agranulocytosis) such as of neutrophil granulocytes (termed neutropenia). Granulocyte deficiencies also include decreased function of individual granulocytes, such as in chronic granulomatous disease.

- Asplenia, where there is no function of the spleen.
- Complement deficiency is where the function of the complement system is deficient.

In reality, immunodeficiency often affects multiple components, with notable examples including severe combined immunodeficiency (which is primary) and acquired immune deficiency syndrome (which is secondary).

Comparison of immunodeficiencies by affected component			
	Affected components	Main causes	Main pathogens of resultant infections
Humoral immune deficiency	B cells, plasma cells or antibodies	• Primary humoral • Multiple myeloma • Chronic lymphoid leukemia • AIDS	• Streptococcus pneumoniae • Hemophilus influenzae • Pneumocystis jirovecii • Giardia intestinalis • Cryptosporidium parvum
T cell deficiency	T cells	• Marrow and other transplantation • AIDS • Cancer chemotherapy • Lymphoma • Glucocorticoid therapy	Intracellular pathogens, including *Herpes simplex virus*, *Mycobacterium*, *Listeria*, and intracellular fungal infections.
Neutropenia	Neutrophil granulocytes	• Chemotherapy • Bone marrow transplantation • Dysfunction, such as chronic granulomatous disease	• Enterobacteriaceae • Oral *Streptococci* • *Pseudomonas aeruginosa* • *Enterococcus* species • *Candida* species • *Aspergillus* species

Asplenia	Spleen	• Splenectomy
		• Trauma
		• Sickle-cell anemia
			• Polysaccharide encapsulated bacteria, particularly:
			  ○ *Streptococcus pneumoniae*
			  ○ *Haemophilus influenzae*
			  ○ *Neisseria meningitidis*
			• *Plasmodium* species
			• *Babesia* species

Complement deficiency	Complement system	• Congenital deficiencies	• *Neisseria* species • Streptococcus pneumoniae

Primary or Secondary

Distinction between primary versus secondary immunodeficiencies are based on, respectively, whether the cause originates in the immune system itself or is, in turn, due to insufficiency of a supporting component of it or an external decreasing factor of it.

Primary Immunodeficiency

A number of rare diseases feature a heightened susceptibility to infections from childhood onward. Primary Immunodeficiency is also known as congenital immunodeficiencies. Many of these disorders are hereditary and are autosomal recessive or X-linked. There are over 80 recognised primary immunodeficiency syndromes; they are generally grouped by the part of the immune system that is malfunctioning, such as lymphocytes or granulocytes.

The treatment of primary immunodeficiencies depends on the nature of the defect, and may involve antibody infusions, long-term antibiotics and (in some cases) stem cell transplantation.

Secondary Immunodeficiencies

Secondary immunodeficiencies, also known as acquired immunodeficiencies, can result from various immunosuppressive agents, for example, malnutrition, aging and particular medications (e.g., chemotherapy, disease-modifying antirheumatic drugs, immunosuppressive drugs after organ transplants, glucocorticoids). For medications, the term immunosuppression generally refers to both beneficial and potential adverse effects of decreasing the function of the immune system, while the term *immunodeficiency* generally refers solely to the adverse effect of increased risk for infection.

Many specific diseases directly or indirectly cause immunosuppression. This includes many types of cancer, particularly those of the bone marrow and blood cells (leukemia, lymphoma, multiple myeloma), and certain chronic infections. Immunodeficiency is also the hallmark of acquired immunodeficiency syndrome (AIDS), caused by the human immunodeficiency virus (HIV). HIV directly infects a small number of T helper cells, and also impairs other immune system responses indirectly.

Immunodeficiency and Autoimmunity

There are a large number of immunodeficiency syndromes that present clinical and laboratory characteristics of autoimmunity. The decreased ability of the immune system to clear infections in these patients may be responsible for causing autoimmunity through perpetual immune system activation.

One example is common variable immunodeficiency (CVID) where multiple autoimmune diseases are seen, e.g., inflammatory bowel disease, autoimmune thrombocytopenia and autoimmune thyroid disease. Familial hemophagocytic lymphohistiocytosis, an autosomal recessive primary immunodeficiency, is another example. Pancytopenia, rashes, lymphadenopathy and hepatosplenomegaly are commonly seen in these patients. Presence of multiple uncleared viral infections due to lack of perforin are thought to be responsible. In addition to chronic and/or recurrent infections many autoimmune diseases including arthritis, autoimmune hemolytic anemia, scleroderma and type 1 diabetes are also seen in X-linked agammaglobulinemia (XLA). Recurrent bacterial and fungal infections and chronic inflammation of the gut and lungs are seen in chronic granulomatous disease (CGD) as well. CGD is caused by a decreased production of nicotinamide adenine dinucleotide phosphate (NADPH) oxidase by neutrophils. Hypomorphic RAG mutations are seen in patients with midline granulomatous disease; an autoimmune disorder that is commonly seen in patients with granulomatosis with polyangiitis (Wegner's disease) and NK/T cell lymphomas. Wiskott-Aldrich syndrome (WAS) patients also present with eczema, autoimmune manifestations, recurrent bacterial infections and lymphoma. In autoimmune polyendocrinopathy-candidiasis-ectodermal dystrophy (APECED) also autoimmunity and infections coexist: organ-specific autoimmune manifestations (e.g., hypoparathyroidism and adrenocortical failure) and chronic mucocutaneous candidiasis. Finally, IgA deficiency is also sometimes associated with the development of autoimmune and atopic phenomena.

Management

Prevention of Pneumocystis pneumonia using trimethoprim/sulfamethoxazole is useful in those who are immunocompromised.

Transplant Rejection

Transplant rejection occurs when transplanted tissue is rejected by the recipient's immune system, which destroys the transplanted tissue. Transplant rejection can be lessened by determining the molecular similitude between donor and recipient and by use of immunosuppressant drugs after transplant.

Pretransplant Rejection Prevention

The first successful organ transplant, performed in 1954 by Joseph Murray, involved identical twins, and so no rejection was observed. Otherwise, the number of mismatched gene variants, namely alleles, encoding cell surface molecules called major histocompatibility complex (MHC), classes I and II, correlate with the rapidity and severity of transplant rejection. In humans MHC is also called human leukocyte antigen (HLA).

Though cytotoxic-crossmatch assay can predict rejection mediated by cellular immunity, genetic-expression tests specific to the organ type to be transplanted, for instance AlloMap Molecular Expression Testing, have a high negative predictive value. Transplanting only ABO-compatible grafts (matching blood groups between donor and recipient) helps prevent rejection mediated by humoral immunity.

ABO-Incompatible Transplants

Because very young children (generally under 12 months, but often as old as 24 months) do not have a well-developed immune system, it is possible for them to receive organs from otherwise incompatible donors. This is known as ABO-incompatible (ABOi) transplantation. Graft survival and patient mortality is approximately the same between ABOi and ABO-compatible (ABOc) recipients. While focus has been on infant heart transplants, the principles generally apply to other forms of solid organ transplantation.

The most important factors are that the recipient not have produced isohemagglutinins, and that they have low levels of T cell-independent antigens. UNOS regulations allow for ABOi transplantation in children under two years of age if isohemagglutinin titers are 1:4 or below, and if there is no matching ABOc recipient. Studies have shown that the period under which a recipient may undergo ABOi transplantation may be prolonged by exposure to nonself A and B antigens. Furthermore, should the recipient (for example, type B-positive with a type AB-positive graft) require eventual retransplantation, the recipient may receive a new organ of either blood type.

Limited success has been achieved in ABO-incompatible heart transplants in adults, though this requires that the adult recipients have low levels of anti-A or anti-B antibodies. Kidney transplantation is more successful, with similar long-term graft survival rates to ABOc transplants.

Immunologic Mechanisms of Rejection

Rejection is an adaptive immune response via cellular immunity (mediated by killer T cells inducing apoptosis of target cells) as well as humoral immunity (mediated by activated B cells secreting antibody molecules), though the action is joined by components of innate immune response (phagocytes and soluble immune proteins). Different types of transplanted tissues tend to favor different balances of rejection mechanisms.

Immunization

An animal's exposure to the antigens of a different member of the same or similar species is *allostimulation*, and the tissue is *allogenic*. Transplanted organs are often acquired from a cadaver (usually a host who had succumbed to trauma), whose tissues had already sustained ischemia or inflammation.

Dendritic cells (DCs), which are the primary antigen-presenting cells (APCs), of the donor tissue migrate to the recipient's peripheral lymphoid tissue (lymphoid follicles and lymph nodes), and present the donor's *self* peptides to the recipient's lymphocytes (immune cells residing in lymphoid tissues). Lymphocytes include two classes that enact adaptive immunity, also called specific immunity. Lymphocytes of specific immunity T cells—including the subclasses helper T cells and killer T cells—and B cells.

The recipient's helper T cells coordinate specific immunity directed at the donor's *self* peptides or at the donor's Major histocompatibility complex molecules, or at both.

Immune Memory

When memory helper T cells' CD4 receptors bind to the MHC class II molecules which are expressed on the surfaces of the target cells of the graft tissue, the memory helper T cells' T cell receptors (TCRs) can recognize their target antigen that is presented by the MHC class II molecules. The memory helper T cell subsequently produces clones that, as effector cells, secrete immune signalling molecules (cytokines) in approximately the cytokine balance that had prevailed at the memory helper T cell's priming to memorize the antigen. As the priming event in this instance occurred amid inflammation, the immune memory is pro-inflammatory.

Cellular Immunity

As a cell is indicated by the prefix *cyto*, a cytotoxic influence destroys the cell. Alloreactive killer T cells, also called cytotoxic T lymphocytes (CTLs), have CD8 receptors that dock to the transplanted tissue's MHC class I molecules,which display the donor's self peptides. (In the living donor, such presentation of *self* antigens helped maintain *self* tolerance.) Thereupon, the T cell receptors (TCRs) of the killer T cells recognize their matching epitope, and trigger the target cell's programmed cell death by apoptosis.

Humoral Immunity

Developed through an earlier *primary exposure* that primed specific immunity to the *nonself* antigen, a transplant recipient can have specific antibody crossreacting with the donor tissue upon the transplant event, a *secondary exposure*. This is typical after earlier mismatching among A/B/O blood types during blood transfusion. At this secondary exposure, these crossreactive antibody molecules interact with aspects of innate immunity—soluble immune proteins called complement and innate immune cells called phagocytes—which inflames and destroys the transplanted tissue.

Antibody

Secreted by an activated B cell, then called plasma cell, an antibody molecule is a soluble immunoglobulin (Ig) whose basic unit is shaped like the letter *Y*: the two arms are the Fab regions, while the single stalk is the Fc region. Each of the two tips of Fab region is the paratope, which binds a matching molecular sequence and its 3D shape (conformation), altogether called epitope, within the target antigen.

Opsonization

The IgG's Fc region also enables opsonization by a phagocyte, a process by which the Fc receptor on the phagocyte—such as neutrophils in blood and macrophages in tissues—binds the antibody molecule's FC stalk, and the phagocyte exhibits enhanced uptake of the antigen, attached to the antibody molecule's Fab region.

Complement Cascade

When the paratope of Ig class *gamma* (IgG) binds its matching epitope, IgG's Fc region conformationally shifts and can host a complement protein, initiating the complement cascade that terminates by punching a hole in a cell membrane. With many holes so punched, fluid rushes into the cell and ruptures it.

Cell debris can be recognized as damage associated molecular patterns (DAMPs) by pattern recognition receptors (PRRs), such as Toll-like receptors (TLRs), on membranes of phagocytes, which thereupon secrete proinflammatory cytokines, recruiting more phagocytes to traffic to the area by sensing the concentration gradient of the secreted cytokines (chemotaxis).

Tissue	Mechanism
Blood	Antibodies (isohaemagglutinins)
Kidney	Antibodies, cell-mediated immunity (CMI)
Heart	Antibodies, CMI

Skin	CMI
Bonemarrow	CMI
Cornea	Usually accepted unless vascularised: CMI

Medical Categories of Rejection

Hyperacute Rejection

Initiated by preexisting humoral immunity, *hyperacute rejection* manifests within minutes after transplant, and if tissue is left implanted brings systemic inflammatory response syndrome. Of high risk in kidney transplants is rapid clumping, namely agglutination, of red blood cells (RBCs or erythrocytes), as an antibody molecule binds multiple target cells at once.

While kidneys can routinely be obtained from human donors, most organs are in short supply leading to consideration of xenotransplants from other species. Pigs are especially likely sources for xenotransplants, chosen for the anatomical and physiological characteristics they share with humans. However, the sugar galactose-alpha-1,3-galactose (αGal) has been implicated as a major factor in hyperacute rejection in xenotransplantation. Unlike virtually all other mammals, humans and other primates do not make αGal, and in fact recognize it as an antigen. During transplantation, xenoreactive natural antibodies recognize αGal on the graft endothelium as an antigen, and the resulting complement-mediated immune response leads to a rejection of the transplant.

Acute Rejection

Developing with formation of cellular immunity, *acute rejection* occurs to some degree in all transplants, except between identical twins, unless immunosuppression is achieved (usually through drugs). Acute rejection begins as early as one week after transplant, the risk being highest in the first three months, though it can occur months to years later. Highly vascular tissues such as kidney or liver often host the earliest signs—particularly at endothelial cells lining blood vessels—though it eventually occurs in roughly 10 to 30% of liver transplants, and 10 to 20% of kidney transplants. A single episode of acute rejection can be recognized and promptly treated, usually preventing organ failure, but recurrent episodes lead to *chronic rejection*. It is believed that the process of acute rejection is mediated by the cell mediated pathway, specifically by mononuclear macrophages and T-lymphocytes.

Chronic Rejection

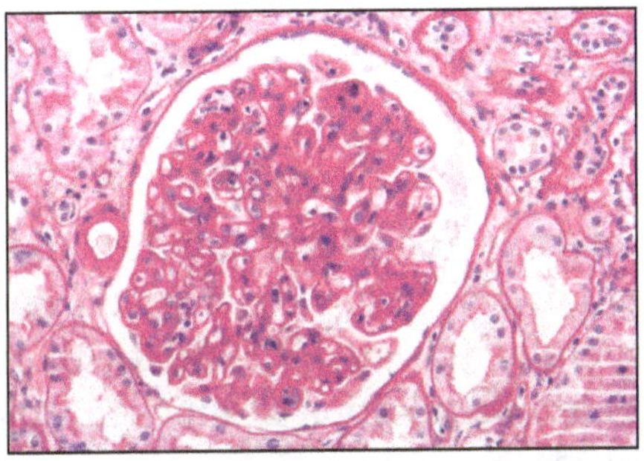

Micrograph showing a glomerulus with changes characteristic of a transplant glomerulopathy. Transplant glomerulopathy is considered a form of chronic antibody-mediated rejection. PAS stain.

The term *chronic rejection* initially described long-term loss of function in transplanted organs via fibrosis of the transplanted tissue's blood vessels. This is now *chronic allograft vasculopathy*, however, leaving *chronic rejection* referring to rejection due to more patent aspects of immunity.

Chronic rejection explains long-term morbidity in most lung-transplant recipients, the median survival roughly 4.7 years, about half the span versus other major organ transplants. In histopathology the condition is *bronchiolitis obliterans*, which clinically presents as progressive airflow obstruction, often involving dyspnea and coughing, and the patient eventually succumbs to pulmonary insufficiency or secondary acute infection.

Airflow obstruction not ascribable to other cause is labeled bronchiolitis obliterans syndrome (BOS), confirmed by a persistent drop—three or more weeks—in *forced expiratory volume* (FEV_1) by at least 20%. BOS is seen in over 50% of lung-transplant recipients by 5 years, and in over 80% by ten years. First noted is infiltration by lymphocytes, followed by epithelial cell injury, then inflammatory lesions and recruitment of fibroblasts and myofibroblasts, which proliferate and secrete proteins forming scar tissue. Generally thought unpredictable, BOS progression varies widely: lung function may suddenly fall but stabilize for years, or rapidly progress to death within a few months. Risk factors include prior acute rejection episodes, gastroesophageal reflux disease, acute infections, particular age groups, HLA mis-matching, lymphocytic bronchiolitis, and graft dysfunction (e.g., airway ischemia).

Rejection due to Non-Adherence

One principal reason for transplant rejection is non-adherence to prescribed immunosuppressant regimens. This is particularly the case with adolescent recipients, with non-adherence rates near 50% in some instances.

Rejection Detection

Diagnosis of acute rejection relies on clinical data—patient signs and symptoms—but also calls on laboratory data such as tissue biopsy. The laboratory pathologist generally seeks three main histological signs: (1) infiltrating T cells, perhaps accompanied by infiltrating eosinophils, plasma cells, and neutrophils, particularly in telltale ratios, (2) structural compromise of tissue anatomy, varying by tissue type transplanted, and (3) injury to blood vessels. Tissue biopsy is restricted, however, by sampling limitations and risks/complications of the invasive procedure. Cellular magnetic resonance imaging (MRI) of immune cells radiolabeled *in vivo* might offer noninvasive testing.

Rejection Treatment

Hyperacute rejection manifests severely and within minutes, and so treatment is immediate: removal of the tissue. Chronic rejection is generally considered irreversible and poorly amenable to treatment—only retransplant generally indicated if feasible—though inhaled ciclosporin is being investigated to delay or prevent chronic rejection of lung transplants. Acute rejection is treated with one or several of a few strategies.

Immunosuppressive Therapy

A short course of high-dose corticosteroids can be applied, and repeated. *Triple therapy* adds a calcineurin inhibitor and an anti-proliferative agent. Where calcineurin inhibitors or steroids are contraindicated, mTOR inhibitors are used.

Immunosuppressive Drugs:

- Corticosteroids:
  - Prednisolone.
  - Hydrocortisone.
- Calcineurin inhibitors:
  - Ciclosporin.
  - Tacrolimus.
- Anti-proliferatives:
  - Azathioprine.
  - Mycophenolic acid.
- mTOR inhibitors:
  - Sirolimus.
  - Everolimus.

Antibody-Based Treatments

Antibody specific to select immune components can be added to immunosuppressive therapy. The monoclonal anti-T cell antibody OKT3, once used to prevent rejection, and still occasionally used to treat severe acute rejection, has fallen into disfavor, as it commonly brings severe cytokine release syndrome and late post-transplant lymphoproliferative disorder. (OKT3 is available in the United Kingdom for named-patient use only.)

Antibody Drugs:

- Monoclonal anti-IL-2Rα receptor antibodies:
  - Basiliximab.
  - Daclizumab.
- Polyclonal anti-T-cell antibodies:
  - Anti-thymocyte globulin (ATG).
  - Anti-lymphocyte globulin (ALG).

- Monoclonal anti-CD20 antibodies:
    - Rituximab.

Blood Transfer

Cases refractory to immunosuppressive or antibody therapy are sometimes given blood transfusions—removing antibody molecules specific to the transplanted tissue.

Marrow Transplant

Bone marrow transplant can replace the transplant recipient's immune system with the donor's, and the recipient accepts the new organ without rejection. The marrow's hematopoietic stem cells—the reservoir of stem cells replenishing exhausted blood cells including white blood cells forming the immune system—must be of the individual who donated the organ or of an identical twin or a clone. There is a risk of graft-versus-host disease (GVHD), however, whereby mature lymphocytes entering with marrow recognize the new host tissues as foreign and destroy them.

Autoimmunity

Autoimmunity is the system of immune responses of an organism against its own healthy cells and tissues. Any disease that results from such an aberrant immune response is termed an autoimmune disease. Prominent examples include celiac disease, diabetes mellitus type 1, sarcoidosis, systemic lupus erythematosus (SLE), Sjögren's syndrome, eosinophilic granulomatosis with polyangiitis, Hashimoto's thyroiditis, Graves' disease, idiopathic thrombocytopenic purpura, Addison's disease, rheumatoid arthritis (RA), ankylosing spondylitis, polymyositis (PM), and dermatomyositis (DM). Autoimmune diseases are very often treated with steroids.

The misconception that an individual's immune system is totally incapable of recognizing *self* antigens is not new. Paul Ehrlich, at the beginning of the twentieth century, proposed the concept of *horror autotoxicus*, wherein a "normal" body does not mount an immune response against its own tissues. Thus, any autoimmune response was perceived to be abnormal and postulated to be connected with human disease. Now, it is accepted that autoimmune responses are an integral part of vertebrate immune systems (sometimes termed "natural autoimmunity"), normally prevented from causing disease by the phenomenon of immunological tolerance to self-antigens. Autoimmunity should not be confused with alloimmunity.

Low-Level Autoimmunity

While a high level of autoimmunity is unhealthy, a low level of autoimmunity may actually be beneficial. Taking the experience of a beneficial factor in autoimmunity further, one might hypothesize with intent to prove that autoimmunity is always a self-defense mechanism of the mammal system to survive. The system does not randomly lose the ability to distinguish between self and non-self, the attack on cells may be the consequence of cycling metabolic processes necessary to keep the blood chemistry in homeostasis.

Second, autoimmunity may have a role in allowing a rapid immune response in the early stages of an infection when the availability of foreign antigens limits the response (i.e., when there are few pathogens present). In their study, Stefanova et al. (2002) injected an anti-MHC Class II antibody into mice expressing a single type of MHC Class II molecule ($H\text{-}2^b$) to temporarily prevent CD4+ T cell-MHC interaction. Naive CD4+ T cells (those that have not encountered any antigens before) recovered from these mice 36 hours post-anti-MHC administration showed decreased responsiveness to the antigen pigeon cytochrome C peptide, as determined by Zap-70 phosphorylation, proliferation, and Interleukin-2 production. Thus Stefanova et al. (2002) demonstrated that self-MHC recognition (which, if too strong may contribute to autoimmune disease) maintains the responsiveness of CD4+ T cells when foreign antigens are absent. This idea of autoimmunity is conceptually similar to play-fighting. The play-fighting of young cubs (TCR and self-MHC) may result in a few scratches or scars (low-level-autoimmunity), but is beneficial in the long-term as it primes the young cub for proper fights in the future.

Immunological Tolerance

Pioneering work by Noel Rose and Ernst Witebsky in New York, and Roitt and Doniach at University College London provided clear evidence that, at least in terms of antibody-producing B lymphocytes, diseases such as rheumatoid arthritis and thyrotoxicosis are associated with loss of immunological tolerance, which is the ability of an individual to ignore "self", while reacting to "non-self". This breakage leads to the immune system's mounting an effective and specific immune response against self determinants. The exact genesis of immunological tolerance is still elusive, but several theories have been proposed since the mid-twentieth century to explain its origin.

Three hypotheses have gained widespread attention among immunologists:

- Clonal Deletion theory, proposed by Burnet, according to which self-reactive lymphoid cells are destroyed during the development of the immune system in an individual. For their work Frank M. Burnet and Peter B. Medawar were awarded the 1960 Nobel Prize in Physiology or Medicine "for discovery of acquired immunological tolerance".
- Clonal Anergy theory, proposed by Nossal, in which self-reactive T- or B-cells become inactivated in the normal individual and cannot amplify the immune response.
- Idiotype Network theory, proposed by Jerne, wherein a network of antibodies capable of neutralizing self-reactive antibodies exists naturally within the body.

In addition, two other theories are under intense investigation:

- Clonal Ignorance theory, according to which autoreactive T cells that are not represented in the thymus will mature and migrate to the periphery, where they will not encounter the appropriate antigen because it is inaccessible tissues. Consequently, auto-reactive B cells, that escape deletion, cannot find the antigen or the specific helper T-cell.
- Suppressor population or Regulatory T cell theory, wherein regulatory T-lymphocytes (commonly $CD4^+FoxP3^+$ cells, among others) function to prevent, downregulate, or limit autoaggressive immune responses in the immune system.

Tolerance can also be differentiated into "Central" and "Peripheral" tolerance, on whether or not the above-stated checking mechanisms operate in the central lymphoid organs (Thymus and Bone Marrow) or the peripheral lymphoid organs (lymph node, spleen, etc., where self-reactive B-cells may be destroyed). It must be emphasised that these theories are not mutually exclusive, and evidence has been mounting suggesting that all of these mechanisms may actively contribute to vertebrate immunological tolerance.

A puzzling feature of the documented loss of tolerance seen in spontaneous human autoimmunity is that it is almost entirely restricted to the autoantibody responses produced by B lymphocytes. Loss of tolerance by T cells has been extremely hard to demonstrate, and where there is evidence for an abnormal T cell response it is usually not to the antigen recognised by autoantibodies. Thus, in rheumatoid arthritis there are autoantibodies to IgG Fc but apparently no corresponding T cell response. In systemic lupus there are autoantibodies to DNA, which cannot evoke a T cell response, and limited evidence for T cell responses implicates nucleoprotein antigens. In Celiac disease there are autoantibodies to tissue transglutaminase but the T cell response is to the foreign protein gliadin. This disparity has led to the idea that human autoimmune disease is in most cases (with probable exceptions including type I diabetes) based on a loss of B cell tolerance which makes use of normal T cell responses to foreign antigens in a variety of aberrant ways.

Immunodeficiency and Autoimmunity

There are a large number of immunodeficiency syndromes that present clinical and laboratory characteristics of autoimmunity. The decreased ability of the immune system to clear infections in these patients may be responsible for causing autoimmunity through perpetual immune system activation.

One example is common variable immunodeficiency (CVID) where multiple autoimmune diseases are seen, e.g. inflammatory bowel disease, autoimmune thrombocytopenia and autoimmune thyroid disease. Familial hemophagocytic lymphohistiocytosis, an autosomal recessive primary immunodeficiency, is another example. Pancytopenia, rashes, swollen lymph nodes and enlargement of the liver and spleen are commonly seen in such individuals. Presence of multiple uncleared viral infections due to lack of perforin are thought to be responsible. In addition to chronic and/or recurrent infections many autoimmune diseases including arthritis, autoimmune hemolytic anemia, scleroderma and type 1 diabetes mellitus are also seen in X-linked agammaglobulinemia (XLA). Recurrent bacterial and fungal infections and chronic inflammation of the gut and lungs are seen in chronic granulomatous disease (CGD) as well. CGD is a caused by decreased production of nicotinamide adenine dinucleotide phosphate (NADPH) oxidase by neutrophils. Hypomorphic RAG mutations are seen in patients with midline granulomatous disease; an autoimmune disorder that is commonly seen in patients with granulomatosis with polyangiitis (formerly known as Wegener's granulomatosis) and NK/T cell lymphomas. Wiskott-Aldrich syndrome (WAS) patients also present with eczema, autoimmune manifestations, recurrent bacterial infections and lymphoma. In autoimmune polyendocrinopathy-candidiasis-ectodermal dystrophy (APECED) also autoimmunity and infections coexist: organ-specific autoimmune manifestations (e.g. hypoparathyroidism and adrenocortical failure) and chronic mucocutaneous candidiasis. Finally, IgA deficiency is also sometimes associated with the development of autoimmune and atopic phenomena.

Genetic Factors

Certain individuals are genetically susceptible to developing autoimmune diseases. This susceptibility is associated with multiple genes plus other risk factors. Genetically predisposed individuals do not always develop autoimmune diseases.

Three main sets of genes are suspected in many autoimmune diseases. These genes are related to:

- Immunoglobulins:
- T-cell receptors:
- The major histocompatibility complexes (MHC).

The first two, which are involved in the recognition of antigens, are inherently variable and susceptible to recombination. These variations enable the immune system to respond to a very wide variety of invaders, but may also give rise to lymphocytes capable of self-reactivity.

Scientists such as Hugh McDevitt, G. Nepom, J. Bell and J. Todd have also provided strong evidence to suggest that certain MHC class II allotypes are strongly correlated with

- HLA DR2 is strongly positively correlated with Systemic Lupus Erythematosus, narcolepsy and multiple sclerosis, and negatively correlated with DM Type 1.
- HLA DR3 is correlated strongly with Sjögren's syndrome, myasthenia gravis, SLE, and DM Type 1.
- HLA DR4 is correlated with the genesis of rheumatoid arthritis, Type 1 diabetes mellitus, and pemphigus vulgaris.

Fewer correlations exist with MHC class I molecules. The most notable and consistent is the association between HLA B27 and spondyloarthropathies like ankylosing spondylitis and reactive arthritis. Correlations may exist between polymorphisms within class II MHC promoters and autoimmune disease.

The contributions of genes outside the MHC complex remain the subject of research, in animal models of disease (Linda Wicker's extensive genetic studies of diabetes in the NOD mouse), and in patients (Brian Kotzin's linkage analysis of susceptibility to SLE).

Recently, PTPN22 has been associated with multiple autoimmune diseases including Type I diabetes, rheumatoid arthritis, systemic lupus erythematosus, Hashimoto's thyroiditis, Graves' disease, Addison's disease, Myasthenia Gravis, vitiligo, systemic sclerosis juvenile idiopathic arthritis, and psoriatic arthritis.

Sex

Ratio of female/male incidence of autoimmune diseases	
Hashimoto's thyroiditis	10/1
Graves' disease	7/1

Multiple sclerosis (MS)	2/1
Myasthenia gravis	2/1
Systemic lupus erythematosus (SLE)	9/1
Rheumatoid arthritis	5/2
Primary sclerosing cholangitis	1/2

A person's sex also seems to have some role in the development of autoimmunity; that is, most autoimmune diseases are *sex-related*. Nearly 75% of the more than 23.5 million Americans who suffer from autoimmune disease are women, although it is less-frequently acknowledged that millions of men also suffer from these diseases. According to the American Autoimmune Related Diseases Association (AARDA), autoimmune diseases that develop in men tend to be more severe. A few autoimmune diseases that men are just as or more likely to develop as women include: ankylosing spondylitis, type 1 diabetes mellitus, granulomatosis with polyangiitis, Crohn's disease, Primary sclerosing cholangitis and psoriasis.

The reasons for the sex role in autoimmunity are unclear. Women appear to generally mount larger inflammatory responses than men when their immune systems are triggered, increasing the risk of autoimmunity. Involvement of sex steroids is indicated by that many autoimmune diseases tend to fluctuate in accordance with hormonal changes, for example, during pregnancy, in the menstrual cycle, or when using oral contraception. A history of pregnancy also appears to leave a persistent increased risk for autoimmune disease. It has been suggested that the slight exchange of cells between mothers and their children during pregnancy may induce autoimmunity. This would tip the gender balance in the direction of the female.

Another theory suggests the female high tendency to get autoimmunity is due to an imbalanced X chromosome inactivation. The X-inactivation skew theory, proposed by Princeton University's Jeff Stewart, has recently been confirmed experimentally in scleroderma and autoimmune thyroiditis. Other complex X-linked genetic susceptibility mechanisms are proposed and under investigation.

Environmental Factors

An interesting inverse relationship exists between infectious diseases and autoimmune diseases. In areas where multiple infectious diseases are endemic, autoimmune diseases are quite rarely seen. The reverse, to some extent, seems to hold true. The hygiene hypothesis attributes these correlations to the immune manipulating strategies of pathogens. Whilst such an observation has been variously termed as spurious and ineffective, according to some studies, parasite infection is associated with reduced activity of autoimmune disease.

The putative mechanism is that the parasite attenuates the host immune response in order to protect itself. This may provide a serendipitous benefit to a host that also suffers from autoimmune disease. The details of parasite immune modulation are not yet known, but may include secretion of anti-inflammatory agents or interference with the host immune signaling.

A paradoxical observation has been the strong association of certain microbial organisms with autoimmune diseases. For example, *Klebsiella pneumoniae* and coxsackievirus B have been strongly correlated with ankylosing spondylitis and diabetes mellitus type 1, respectively. This has been explained by the tendency of the infecting organism to produce super-antigens that are capable of polyclonal activation of B-lymphocytes, and production of large amounts of antibodies of varying specificities, some of which may be self-reactive.

Certain chemical agents and drugs can also be associated with the genesis of autoimmune conditions, or conditions that simulate autoimmune diseases. The most striking of these is the drug-induced lupus erythematosus. Usually, withdrawal of the offending drug cures the symptoms in a patient.

Cigarette smoking is now established as a major risk factor for both incidence and severity of rheumatoid arthritis. This may relate to abnormal citrullination of proteins, since the effects of smoking correlate with the presence of antibodies to citrullinated peptides.

Pathogenesis of Autoimmunity

Several mechanisms are thought to be operative in the pathogenesis of autoimmune diseases, against a backdrop of genetic predisposition and environmental modulation. It is beyond the scope of this article to discuss each of these mechanisms exhaustively, but a summary of some of the important mechanisms have been described:

- T-Cell Bypass – A normal immune system requires the activation of B-cells by T-cells before the former can undergo differentiation into plasma B-cells and subsequently produce antibodies in large quantities. This requirement of a T-cell can be bypassed in rare instances, such as infection by organisms producing super-antigens, which are capable of initiating polyclonal activation of B-cells, or even of T-cells, by directly binding to the β-subunit of T-cell receptors in a non-specific fashion.
- T-Cell-B-Cell discordance – A normal immune response is assumed to involve B and T cell responses to the same antigen, even if we know that B cells and T cells recognise very different things: conformations on the surface of a molecule for B cells and pre-processed peptide fragments of proteins for T cells. However, there is nothing as far as we know that requires this. All that is required is that a B cell recognising antigen X endocytoses and processes a protein Y (normally =X) and presents it to a T cell. Roosnek and Lanzavecchia showed that B cells recognising IgGFc could get help from any T cell responding to an antigen co-endocytosed with IgG by the B cell as part of an immune complex. In coeliac disease it seems likely that B cells recognising tissue transglutamine are helped by T cells recognising gliadin.
- Aberrant B cell receptor-mediated feedback – A feature of human autoimmune disease is that it is largely restricted to a small group of antigens, several of which have known signaling roles in the immune response (DNA, C1q, IgGFc, Ro, Con. A receptor, Peanut agglutinin receptor(PNAR)). This fact gave rise to the idea that spontaneous autoimmunity may result when the binding of antibody to certain antigens leads to aberrant signals being fed back to parent B cells through membrane bound ligands. These ligands include B cell receptor (for antigen), IgG Fc receptors, CD21, which binds complement C3d, Toll-like receptors 9 and 7 (which can bind DNA and nucleoproteins) and PNAR. More indirect aberrant activation of B cells can also be envisaged with autoantibodies to acetyl choline receptor (on thymic myoid cells) and hormone and hormone binding proteins. Together with the concept of T-cell-B-cell discordance this idea forms the basis of the hypothesis of self-perpetuating autoreactive B cells. Autoreactive B cells in spontaneous autoimmunity

are seen as surviving because of subversion both of the T cell help pathway and of the feedback signal through B cell receptor, thereby overcoming the negative signals responsible for B cell self-tolerance without necessarily requiring loss of T cell self-tolerance.

- Molecular Mimicry – An exogenous antigen may share structural similarities with certain host antigens; thus, any antibody produced against this antigen (which mimics the self-antigens) can also, in theory, bind to the host antigens, and amplify the immune response. The idea of molecular mimicry arose in the context of Rheumatic Fever, which follows infection with Group A beta-haemolytic streptococci. Although rheumatic fever has been attributed to molecular mimicry for half a century no antigen has been formally identified (if anything too many have been proposed). Moreover, the complex tissue distribution of the disease (heart, joint, skin, basal ganglia) argues against a cardiac specific antigen. It remains entirely possible that the disease is due to e.g. an unusual interaction between immune complexes, complement components and endothelium.
- Idiotype Cross-Reaction – Idiotypes are antigenic epitopes found in the antigen-binding portion (Fab) of the immunoglobulin molecule. Plotz and Oldstone presented evidence that autoimmunity can arise as a result of a cross-reaction between the idiotype on an antiviral antibody and a host cell receptor for the virus in question. In this case, the host-cell receptor is envisioned as an internal image of the virus, and the anti-idiotype antibodies can react with the host cells.
- Cytokine Dysregulation – Cytokines have been recently divided into two groups according to the population of cells whose functions they promote: Helper T-cells type 1 or type 2. The second category of cytokines, which include IL-4, IL-10 and TGF-β (to name a few), seem to have a role in prevention of exaggeration of pro-inflammatory immune responses.
- Dendritic cell apoptosis – immune system cells called dendritic cells present antigens to active lymphocytes. Dendritic cells that are defective in apoptosis can lead to inappropriate systemic lymphocyte activation and consequent decline in self-tolerance.
- Epitope spreading or epitope drift – when the immune reaction changes from targeting the primary epitope to also targeting other epitopes. In contrast to molecular mimicry, the other epitopes need not be structurally similar to the primary one.
- Epitope modification or Cryptic epitope exposure – this mechanism of autoimmune disease is unique in that it does not result from a defect in the hematopoietic system. Instead, disease results from the exposure of cryptic N-glycan (polysaccharide) linkages common to lower eukaryotes and prokaryotes on the glycoproteins of mammalian non-hematopoietic cells and organs This exposure of phylogenically primitive glycans activates one or more mammalian innate immune cell receptors to induce a chronic sterile inflammatory state. In the presence of chronic and inflammatory cell damage, the adaptive immune system is recruited and self–tolerance is lost with increased autoantibody production. In this form of the disease, the absence of lymphocytes can accelerate organ damage, and intravenous IgG administration can be therapeutic. Although this route to autoimmune disease may underlie various degenerative disease states, no diagnostics for this disease mechanism exist at present, and thus its role in human autoimmunity is currently unknown.

The roles of specialized immunoregulatory cell types, such as regulatory T cells, NKT cells, γδ T-cells in the pathogenesis of autoimmune disease are under investigation.

Classification

Autoimmune diseases can be broadly divided into systemic and organ-specific or localised autoimmune disorders, depending on the principal clinico-pathologic features of each disease.

- Systemic autoimmune diseases include SLE, Sjögren's syndrome, sarcoidosis, scleroderma, rheumatoid arthritis, cryoglobulinemic vasculitis, and dermatomyositis. These conditions tend to be associated with autoantibodies to antigens which are not tissue specific. Thus although polymyositis is more or less tissue specific in presentation, it may be included in this group because the autoantigens are often ubiquitous t-RNA synthetases.
- Local syndromes which affect a specific organ or tissue:
  - Endocrinologic: Diabetes mellitus type 1, Hashimoto's thyroiditis, Addison's disease.
  - Gastrointestinal: Coeliac disease, Crohn's Disease, Pernicious anaemia.
  - Dermatologic: Pemphigus vulgaris, Vitiligo.
  - Haematologic: Autoimmune haemolytic anaemia, Idiopathic thrombocytopenic purpura.
  - Neurological: Myasthenia gravis.

Using the traditional "organ specific" and "non-organ specific" classification scheme, many diseases have been lumped together under the autoimmune disease umbrella. However, many chronic inflammatory human disorders lack the telltale associations of B and T cell driven immunopathology. In the last decade it has been firmly established that tissue "inflammation against self" does not necessarily rely on abnormal T and B cell responses.

This has led to the recent proposal that the spectrum of autoimmunity should be viewed along an "immunological disease continuum," with classical autoimmune diseases at one extreme and diseases driven by the innate immune system at the other extreme. Within this scheme, the full spectrum of autoimmunity can be included. Many common human autoimmune diseases can be seen to have a substantial innate immune mediated immunopathology using this new scheme. This new classification scheme has implications for understanding disease mechanisms and for therapy development.

Diagnosis

Diagnosis of autoimmune disorders largely rests on accurate history and physical examination of the patient, and high index of suspicion against a backdrop of certain abnormalities in routine laboratory tests (example, elevated C-reactive protein). In several systemic disorders, serological assays which can detect specific autoantibodies can be employed. Localised disorders are best diagnosed by immunofluorescence of biopsy specimens. Autoantibodies are used to diagnose many autoimmune diseases. The levels of autoantibodies are measured to determine the progress of the disease.

Treatments

Treatments for autoimmune disease have traditionally been immunosuppressive, anti-inflammatory, or palliative. Managing inflammation is critical in autoimmune diseases. Non-immunological therapies, such as hormone replacement in Hashimoto's thyroiditis or Type 1 diabetes mellitus treat outcomes of the autoaggressive response, thus these are palliative treatments. Dietary manipulation limits the severity of celiac disease. Steroidal or NSAID treatment limits inflammatory symptoms of many diseases. IVIG is used for CIDP and GBS. Specific immunomodulatory therapies, such as the TNFα antagonists (e.g. etanercept), the B cell depleting agent rituximab, the anti-IL-6 receptor tocilizumab and the costimulation blocker abatacept have been shown to be useful in treating RA. Some of these immunotherapies may be associated with increased risk of adverse effects, such as susceptibility to infection.

Helminthic therapy is an experimental approach that involves inoculation of the patient with specific parasitic intestinal nematodes (helminths). There are currently two closely related treatments available, inoculation with either Necator americanus, commonly known as hookworms, or Trichuris Suis Ova, commonly known as Pig Whipworm Eggs.

T cell vaccination is also being explored as a possible future therapy for autoimmune disorders.

Nutrition and Autoimmunity

Vitamin D/Sunlight:

- Because most human cells and tissues have receptors for vitamin D, including T and B cells, adequate levels of vitamin D can aid in the regulation of the immune system.

Omega-3 Fatty Acids:

- Studies have shown that adequate consumption of omega-3 fatty acids counteracts the effects of arachidonic acids, which contribute to symptoms of autoimmune diseases. Human and animal trials suggest that omega-3 is an effective treatment modality for many cases of Rheumatoid Arthritis, Inflammatory Bowel Disease, Asthma, and Psoriasis.
- While major depression is not necessarily an autoimmune disease, some of is physiological symptoms are inflammatory and autoimmune in nature. Omega-3 may inhibit production of interferon gamma and other cytokines which cause the physiological symptoms of depression. This may be due to the fact that an imbalance in omega-3 and omega-6 fatty acids, which have opposing effects, is instrumental in the etiology of major depression.

Probiotics/Microflora:

- Various types of bacteria and microflora present in fermented dairy products, especially *Lactobacillus casei*, have been shown to both stimulate immune response to tumors in mice and to regulate immune function, delaying or preventing the onset of nonobese diabetes. This is particularly true of the Shirota strain of *L. casei* (LcS). The LcS strain is mainly found in yogurt and similar products in Europe and Japan, and rarely elsewhere.

Antioxidants:

- It has been theorized that free radicals contribute to the onset of type-1 diabetes in infants and young children, and therefore that the risk could be reduced by high intake of antioxidant substances during pregnancy. However, a study conducted in a hospital in Finland from 1997-2002 concluded that there was no statistically significant correlation between antioxidant intake and diabetes risk. It should be noted that this study involved monitoring of food intake through questionnaires, and estimated antioxidant intake on this basis, rather than by exact measurements or use of supplements.

HIV/AIDS

Human immunodeficiency virus infection and acquired immune deficiency syndrome (HIV/AIDS) is a spectrum of conditions caused by infection with the human immunodeficiency virus (HIV). Following initial infection, a person may not notice any symptoms or may experience a brief period of influenza-like illness. Typically, this is followed by a prolonged period with no symptoms. As the infection progresses, it interferes more with the immune system, increasing the risk of common infections like tuberculosis, as well as other opportunistic infections, and tumors that rarely affect people who have working immune systems. These late symptoms of infection are referred to as AIDS. This stage is often also associated with weight loss.

HIV is spread primarily by unprotected sex (including anal and oral sex), contaminated blood transfusions, hypodermic needles, and from mother to child during pregnancy, delivery, or breastfeeding. Some bodily fluids, such as saliva and tears, do not transmit HIV. Methods of prevention include safe sex, needle exchange programmes, treating those who are infected, and male circumcision. Disease in a baby can often be prevented by giving both the mother and child antiretroviral medication. There is no cure or vaccine; however, antiretroviral treatment can slow the course of the disease and may lead to a near-normal life expectancy. Treatment is recommended as soon as the diagnosis is made. Without treatment, the average survival time after infection is 11 years.

In 2014 about 36.9 million people were living with HIV and it resulted in 1.2 million deaths. Most of those infected live in sub-Saharan Africa. Between its discovery and 2014 AIDS has caused an estimated 39 million deaths worldwide. HIV/AIDS is considered a pandemic—a disease outbreak which is present over a large area and is actively spreading. HIV is believed to have originated in west-central Africa during the late 19th or early 20th century. AIDS was first recognized by the United States Centers for Disease Control and Prevention (CDC) in 1981 and its cause—HIV infection—was identified in the early part of the decade.

HIV/AIDS has had a great impact on society, both as an illness and as a source of discrimination. The disease also has large economic impacts. There are many misconceptions about HIV/AIDS such as the belief that it can be transmitted by casual non-sexual contact. The disease has become subject to many controversies involving religion including the Catholic church's decision not to support condom use as prevention. It has attracted international medical and political attention as well as large-scale funding since it was identified in the 1980s.

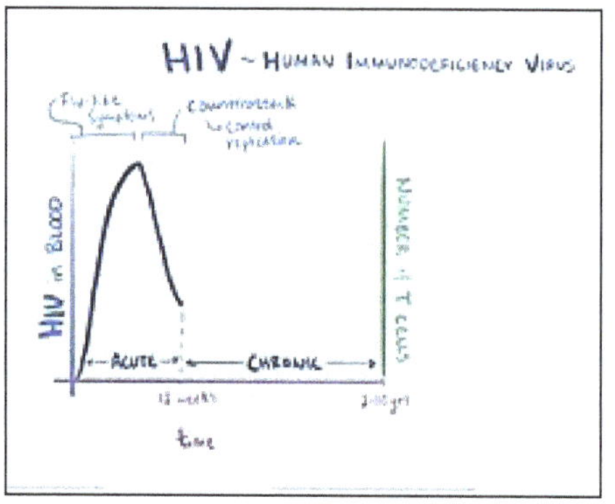

Video explanation.

Signs and Symptoms

There are three main stages of HIV infection: acute infection, clinical latency and AIDS.

Acute Infection

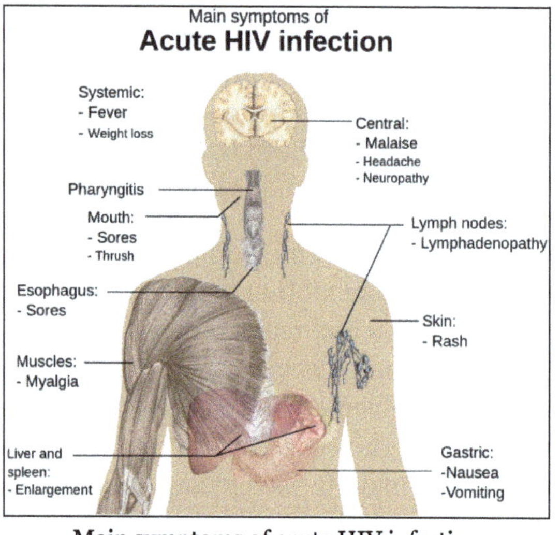

Main symptoms of acute HIV infection.

The initial period following the contraction of HIV is called acute HIV, primary HIV or acute retroviral syndrome. Many individuals develop an influenza-like illness or a mononucleosis-like illness 2–4 weeks post exposure while others have no significant symptoms. Symptoms occur in 40–90% of cases and most commonly include fever, large tender lymph nodes, throat inflammation, a rash, headache, and/or sores of the mouth and genitals. The rash, which occurs in 20–50% of cases, presents itself on the trunk and is maculopapular, classically. Some people also develop opportunistic infections at this stage. Gastrointestinal symptoms such as nausea, vomiting or diarrhea may occur, as may neurological symptoms of peripheral neuropathy or Guillain-Barre syndrome. The duration of the symptoms varies, but is usually one or two weeks.

Due to their nonspecific character, these symptoms are not often recognized as signs of HIV infection. Even cases that do get seen by a family doctor or a hospital are often misdiagnosed as one of the many common infectious diseases with overlapping symptoms. Thus, it is recommended that HIV be considered in people presenting an unexplained fever who may have risk factors for the infection.

Clinical Latency

The initial symptoms are followed by a stage called clinical latency, asymptomatic HIV, or chronic HIV. Without treatment, this second stage of the natural history of HIV infection can last from about three years to over 20 years (on average, about eight years). While typically there are few or no symptoms at first, near the end of this stage many people experience fever, weight loss, gastrointestinal problems and muscle pains. Between 50 and 70% of people also develop persistent generalized lymphadenopathy, characterized by unexplained, non-painful enlargement of more than one group of lymph nodes (other than in the groin) for over three to six months.

Although most HIV-1 infected individuals have a detectable viral load and in the absence of treatment will eventually progress to AIDS, a small proportion (about 5%) retain high levels of $CD4^{+}$ T cells (T helper cells) without antiretroviral therapy for more than 5 years. These individuals are classified as HIV controllers or long-term nonprogressors (LTNP). Another group consists of those who maintain a low or undetectable viral load without anti-retroviral treatment, known as "elite controllers" or "elite suppressors". They represent approximately 1 in 300 infected persons.

Acquired Immunodeficiency Syndrome

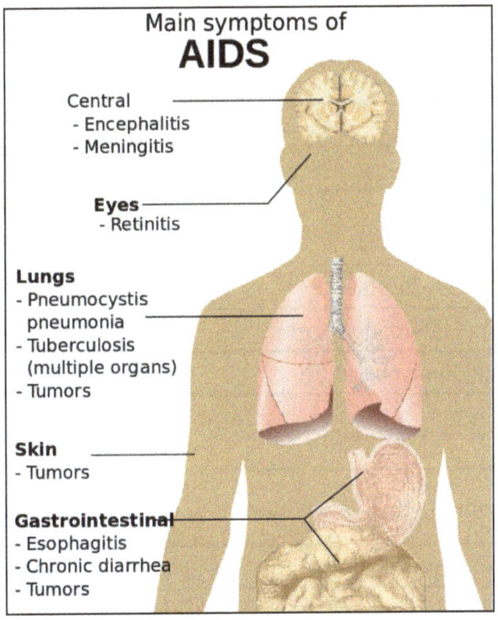

Main symptoms of AIDS.

Acquired immunodeficiency syndrome (AIDS) is defined in terms of either a $CD4^{+}$ T cell count below 200 cells per µL or the occurrence of specific diseases in association with an HIV infection. In the absence of specific treatment, around half of people infected with HIV develop AIDS within ten years. The most common initial conditions that alert to the presence of AIDS are pneumocystis pneumonia (40%), cachexia in the form of HIV wasting syndrome (20%), and esophageal candidiasis. Other common signs include recurring respiratory tract infections.

Opportunistic infections may be caused by bacteria, viruses, fungi, and parasites that are normally controlled by the immune system. Which infections occur depends partly on what organisms are common in the person's environment. These infections may affect nearly every organ system.

People with AIDS have an increased risk of developing various viral-induced cancers, including Kaposi's sarcoma, Burkitt's lymphoma, primary central nervous system lymphoma, and cervical cancer. Kaposi's sarcoma is the most common cancer occurring in 10 to 20% of people with HIV. The second most common cancer is lymphoma, which is the cause of death of nearly 16% of people with AIDS and is the initial sign of AIDS in 3 to 4%. Both these cancers are associated with human herpesvirus 8. Cervical cancer occurs more frequently in those with AIDS because of its association with human papillomavirus (HPV). Conjunctival cancer (of the layer that lines the inner part of eyelids and the white part of the eye) is also more common in those with HIV.

Additionally, people with AIDS frequently have systemic symptoms such as prolonged fevers, sweats (particularly at night), swollen lymph nodes, chills, weakness, and unintended weight loss. Diarrhea is another common symptom, present in about 90% of people with AIDS. They can also be affected by diverse psychiatric and neurological symptoms independent of opportunistic infections and cancers.

Transmission

Average per act risk of getting HIV by exposure route to an infected source	
Exposure route	Chance of infection
Blood transfusion	90%
Childbirth (to child)	25%
Needle-sharing injection drug use	0.67%
Percutaneous needle stick	0.30%
Receptive anal intercourse*	0.04–3.0%
Insertive anal intercourse*	0.03%
Receptive penile-vaginal intercourse*	0.05–0.30%
Insertive penile-vaginal intercourse*	0.01–0.38%
Receptive oral intercourse*§	0–0.04%
Insertive oral intercourse*§	0–0.005%
* assuming no condom use § source refers to oral intercourse performed on a man	

HIV is transmitted by three main routes: sexual contact, significant exposure to infected body fluids or tissues, and from mother to child during pregnancy, delivery, or breastfeeding (known as vertical transmission). There is no risk of acquiring HIV if exposed to feces, nasal secretions, saliva, sputum, sweat, tears, urine, or vomit unless these are contaminated with blood. It is possible to be co-infected by more than one strain of HIV—a condition known as HIV superinfection.

Sexual

The most frequent mode of transmission of HIV is through sexual contact with an infected person. The majority of all transmissions worldwide occur through heterosexual contacts (i.e. sexual contacts between people of the opposite sex); however, the pattern of transmission varies significantly among countries. In the United States, as of 2010, most transmission occurred in men who had sex with men, with this population accounting for 65% of all new cases.

With regard to unprotected heterosexual contacts, estimates of the risk of HIV transmission per sexual act appear to be four to ten times higher in low-income countries than in high-income countries. In low-income countries, the risk of female-to-male transmission is estimated as 0.38% per act, and of male-to-female transmission as 0.30% per act; the equivalent estimates for high-income countries are 0.04% per act for female-to-male transmission, and 0.08% per act for male-to-female transmission. The risk of transmission from anal intercourse is especially high, estimated as 1.4–1.7% per act in both heterosexual and homosexual contacts. While the risk of transmission from oral sex is relatively low, it is still present. The risk from receiving oral sex has been described as "nearly nil"; however, a few cases have been reported. The per-act risk is estimated at 0–0.04% for receptive oral intercourse. In settings involving prostitution in low income countries, risk of female-to-male transmission has been estimated as 2.4% per act and male-to-female transmission as 0.05% per act.

Risk of transmission increases in the presence of many sexually transmitted infections and genital ulcers. Genital ulcers appear to increase the risk approximately fivefold. Other sexually transmitted infections, such as gonorrhea, chlamydia, trichomoniasis, and bacterial vaginosis, are associated with somewhat smaller increases in risk of transmission.

The viral load of an infected person is an important risk factor in both sexual and mother-to-child transmission. During the first 2.5 months of an HIV infection a person's infectiousness is twelve times higher due to this high viral load. If the person is in the late stages of infection, rates of transmission are approximately eightfold greater.

Commercial sex workers (including those in pornography) have an increased rate of HIV. Rough sex can be a factor associated with an increased risk of transmission. Sexual assault is also believed to carry an increased risk of HIV transmission as condoms are rarely worn, physical trauma to the vagina or rectum is likely, and there may be a greater risk of concurrent sexually transmitted infections.

Body Fluids

The second most frequent mode of HIV transmission is via blood and blood products. Blood-borne transmission can be through needle-sharing during intravenous drug use, needle stick injury, transfusion of contaminated blood or blood product, or medical injections with unsterilised equipment. The risk from sharing a needle during drug injection is between 0.63 and 2.4% per act, with an average of 0.8%. The risk of acquiring HIV from a needle stick from an HIV-infected person is estimated as 0.3% (about 1 in 333) per act and the risk following mucous membrane exposure to infected blood as 0.09% (about 1 in 1000) per act. In the United States intravenous drug users made up 12% of all new cases of HIV in 2009, and in some areas more than 80% of people who inject drugs are HIV positive.

HIV is transmitted in about 93% of blood transfusions using infected blood. In developed countries the risk of acquiring HIV from a blood transfusion is extremely low (less than one in half a million) where improved donor selection and HIV screening is performed; for example, in the UK the risk is reported at one in five million and in the United States it was one in 1.5 million in 2008. In low income countries, only half of transfusions may be appropriately screened (as of 2008), and it is estimated that up to 15% of HIV infections in these areas come from transfusion of infected blood and blood products, representing between 5% and 10% of global infections. Although rare because of screening, it is possible to acquire HIV from organ and tissue transplantation.

CDC poster from 1989 highlighting the threat of AIDS associated with drug use.

Unsafe medical injections play a significant role in HIV spread in sub-Saharan Africa. In 2007, between 12 and 17% of infections in this region were attributed to medical syringe use. The World Health Organization estimates the risk of transmission as a result of a medical injection in Africa at 1.2%. Significant risks are also associated with invasive procedures, assisted delivery, and dental care in this area of the world.

People giving or receiving tattoos, piercings, and scarification are theoretically at risk of infection but no confirmed cases have been documented. It is not possible for mosquitoes or other insects to transmit HIV.

Mother-To-Child

HIV can be transmitted from mother to child during pregnancy, during delivery, or through breast milk resulting in infection in the baby. This is the third most common way in which HIV is transmitted globally. In the absence of treatment, the risk of transmission before or during birth is around 20% and in those who also breastfeed 35%. As of 2008, vertical transmission accounted for about 90% of cases of HIV in children. With appropriate treatment the risk of mother-to-child infection can be reduced to about 1%. Preventive treatment involves the mother taking antiretrovirals during pregnancy and delivery, an elective caesarean section, avoiding breastfeeding, and administering antiretroviral drugs to the newborn. Antiretrovirals when taken by either the mother or the infant decrease the risk of transmission in those who do breastfeed. Many of these measures are however not available in the developing world. If blood contaminates food during pre-chewing it may pose a risk of transmission.

Virology

HIV is the cause of the spectrum of disease known as HIV/AIDS. HIV is a retrovirus that primarily infects components of the human immune system such as $CD4^+$ T cells, macrophages and dendritic cells. It directly and indirectly destroys $CD4^+$ T cells.

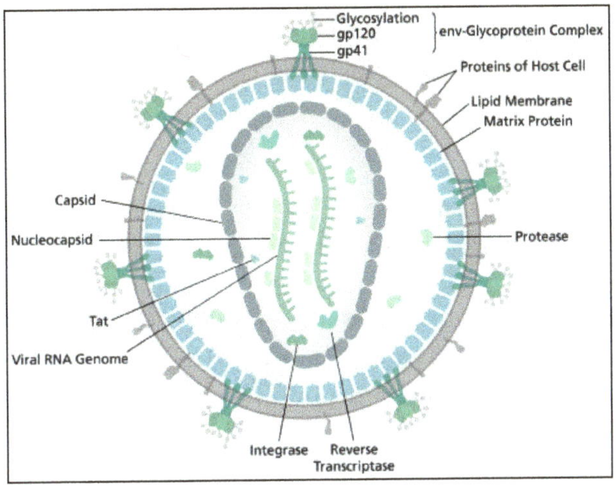

Diagram of a HIV virion structure.

HIV is a member of the genus *Lentivirus*, part of the family *Retroviridae*. Lentiviruses share many morphological and biological characteristics. Many species of mammals are infected by lentiviruses, which are characteristically responsible for long-duration illnesses with a long incubation period. Lentiviruses are transmitted as single-stranded, positive-sense, enveloped RNA viruses. Upon entry into the target cell, the viral RNA genome is converted (reverse transcribed) into double-stranded DNA by a virally encoded reverse transcriptase that is transported along with the viral genome in the virus particle. The resulting viral DNA is then imported into the cell nucleus and integrated into the cellular DNA by a virally encoded integrase and host co-factors. Once integrated, the virus may become latent, allowing the virus and its host cell to avoid detection by the immune system. Alternatively, the virus may be transcribed, producing new RNA genomes and viral proteins that are packaged and released from the cell as new virus particles that begin the replication cycle anew.

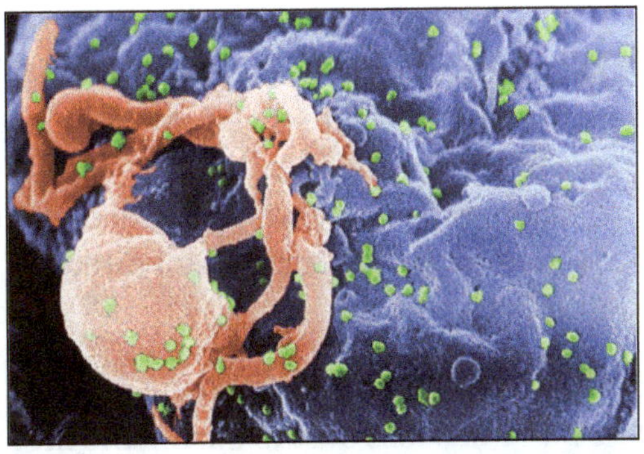

Scanning electron micrograph of HIV-1, colored green, budding from a cultured lymphocyte.

HIV is now known to spread between $CD4^+$ T cells by two parallel routes: cell-free spread and cell-to-cell spread, i.e. it employs hybrid spreading mechanisms. In the cell-free spread, virus particles bud from an infected T cell, enter the blood/extracellular fluid and then infect another T cell following a chance encounter. HIV can also disseminate by direct transmission from one cell to another by a process of cell-to-cell spread. The hybrid spreading mechanisms of HIV contribute to the virus's ongoing replication against antiretroviral therapies.

Two types of HIV have been characterized: HIV-1 and HIV-2. HIV-1 is the virus that was originally discovered (and initially referred to also as LAV or HTLV-III). It is more virulent, more infective, and is the cause of the majority of HIV infections globally. The lower infectivity of HIV-2 as compared with HIV-1 implies that fewer people exposed to HIV-2 will be infected per exposure. Because of its relatively poor capacity for transmission, HIV-2 is largely confined to West Africa.

Pathophysiology

After the virus enters the body there is a period of rapid viral replication, leading to an abundance of virus in the peripheral blood. During primary infection, the level of HIV may reach several million virus particles per milliliter of blood. This response is accompanied by a marked drop in the number of circulating $CD4^+$ T cells. The acute viremia is almost invariably associated with activation of $CD8^+$ T cells, which kill HIV-infected cells, and subsequently with antibody production, or seroconversion. The $CD8^+$ T cell response is thought to be important in controlling virus levels, which peak and then decline, as the $CD4^+$ T cell counts recover. A good $CD8^+$ T cell response has been linked to slower disease progression and a better prognosis, though it does not eliminate the virus.

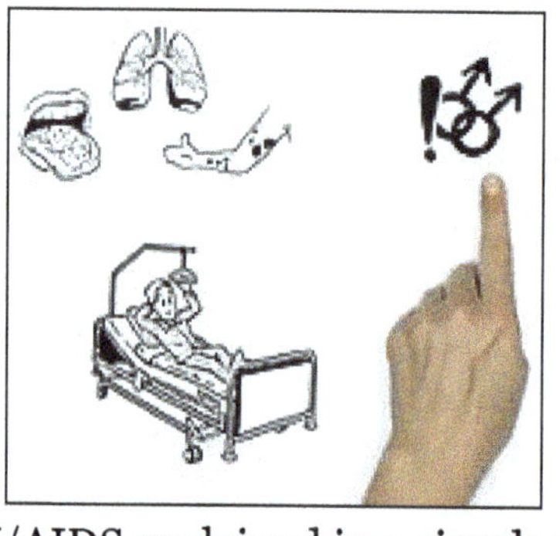

HIV/AIDS explained in a simple way.

Ultimately, HIV causes AIDS by depleting $CD4^+$ T cells. This weakens the immune system and allows opportunistic infections. T cells are essential to the immune response and without them, the body cannot fight infections or kill cancerous cells. The mechanism of $CD4^+$ T cell depletion differs in the acute and chronic phases. During the acute phase, HIV-induced cell lysis and killing of infected cells by cytotoxic T cells accounts for $CD4^+$ T cell depletion, although apoptosis may also be a factor. During the chronic phase, the consequences of generalized immune activation coupled with the gradual loss of the ability of the immune system to generate new T cells appear to account for the slow decline in $CD4^+$ T cell numbers.

Although the symptoms of immune deficiency characteristic of AIDS do not appear for years after a person is infected, the bulk of $CD4^+$ T cell loss occurs during the first weeks of infection, especially in the intestinal mucosa, which harbors the majority of the lymphocytes found in the body. The reason for the preferential loss of mucosal CD4⁺ T cells is that the majority of mucosal CD4⁺ T cells express the CCR5 protein which HIV uses as a co-receptor to gain access to the cells, whereas only a small fraction of CD4⁺ T cells in the bloodstream do so. A specific genetic change that alters the CCR5 protein when present in both chromosomes very effectively prevents HIV-1 infection.

HIV seeks out and destroys CCR5 expressing CD4⁺ T cells during acute infection. A vigorous immune response eventually controls the infection and initiates the clinically latent phase. CD4⁺ T cells in mucosal tissues remain particularly affected. Continuous HIV replication causes a state of generalized immune activation persisting throughout the chronic phase. Immune activation, which is reflected by the increased activation state of immune cells and release of pro-inflammatory cytokines, results from the activity of several HIV gene products and the immune response to ongoing HIV replication. It is also linked to the breakdown of the immune surveillance system of the gastrointestinal mucosal barrier caused by the depletion of mucosal CD4⁺ T cells during the acute phase of disease.

Diagnosis

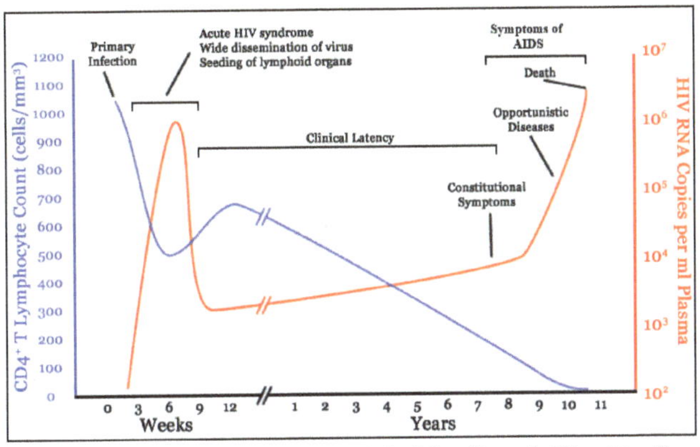

A generalized graph of the relationship between HIV copies (viral load) and CD4⁺ T cell counts over the average course of untreated HIV infection.

- CD4⁺ T Lymphocyte count (cells/mm³).
- HIV RNA copies per mL of plasma.

HIV/AIDS is diagnosed via laboratory testing and then staged based on the presence of certain signs or symptoms. HIV screening is recommended by the United States Preventive Services Task Force for all people 15 years to 65 years of age including all pregnant women. Additionally, testing is recommended for those at high risk, which includes anyone diagnosed with a sexually transmitted illness. In many areas of the world, a third of HIV carriers only discover they are infected at an advanced stage of the disease when AIDS or severe immunodeficiency has become apparent.

HIV Testing

Most people infected with HIV develop specific antibodies (i.e. seroconvert) within three to twelve weeks of the initial infection. Diagnosis of primary HIV before seroconversion is done by measuring HIV-RNA or p24 antigen. Positive results obtained by antibody or PCR testing are confirmed either by a different antibody or by PCR.

Antibody tests in children younger than 18 months are typically inaccurate due to the continued presence of maternal antibodies. Thus HIV infection can only be diagnosed by PCR testing for HIV RNA or DNA, or via testing for the p24 antigen. Much of the world lacks access to reliable PCR testing and many places simply wait until either symptoms develop or the child is old enough for accurate antibody testing. In sub-Saharan Africa as of 2007–2009 between 30 and 70% of the population were aware of their HIV status. In 2009, between 3.6 and 42% of men and women in Sub-Saharan countries were tested which represented a significant increase compared to previous years.

Classifications

Two main clinical staging systems are used to classify HIV and HIV-related disease for surveillance purposes: the WHO disease staging system for HIV infection and disease, and the CDC classification system for HIV infection. The CDC's classification system is more frequently adopted in developed countries. Since the WHO's staging system does not require laboratory tests, it is suited to the resource-restricted conditions encountered in developing countries, where it can also be used to help guide clinical management. Despite their differences, the two systems allow comparison for statistical purposes.

The World Health Organization first proposed a definition for AIDS in 1986. Since then, the WHO classification has been updated and expanded several times, with the most recent version being published in 2007. The WHO system uses the following categories:

- Primary HIV infection: May be either asymptomatic or associated with acute retroviral syndrome.
- Stage I: HIV infection is asymptomatic with a $CD4^+$ T cell count (also known as CD4 count) greater than 500 per microlitre (µl or cubic mm) of blood. May include generalized lymph node enlargement.
- Stage II: Mild symptoms which may include minor mucocutaneous manifestations and recurrent upper respiratory tract infections. A CD4 count of less than 500/µl.
- Stage III: Advanced symptoms which may include unexplained chronic diarrhea for longer than a month, severe bacterial infections including tuberculosis of the lung, and a CD4 count of less than 350/µl.
- Stage IV or AIDS: severe symptoms which include toxoplasmosis of the brain, candidiasis of the esophagus, trachea, bronchi or lungs and Kaposi's sarcoma. A CD4 count of less than 200/µl.

The United States Center for Disease Control and Prevention also created a classification system for HIV, and updated it in 2008 and 2014. This system classifies HIV infections based on CD4 count and clinical symptoms, and describes the infection in five groups. In those greater than six years of age it is:

- Stage 0: the time between a negative or indeterminate HIV test followed less than 180 days by a positive test.
- Stage 1: CD4 count ≥ 500 cells/µl and no AIDS defining conditions.

- Stage 2: CD4 count 200 to 500 cells/μl and no AIDS defining conditions.
- Stage 3: CD4 count ≤ 200 cells/μl or AIDS defining conditions.
- Unknown: if insufficient information is available to make any of the above classifications.

For surveillance purposes, the AIDS diagnosis still stands even if, after treatment, the $CD4^+$ T cell count rises to above 200 per μL of blood or other AIDS-defining illnesses are cured.

Prevention

AIDS Clinic, McLeod Ganj, Himachal Pradesh, India, 2010.

Sexual Contact

Consistent condom use reduces the risk of HIV transmission by approximately 80% over the long term. When condoms are used consistently by a couple in which one person is infected, the rate of HIV infection is less than 1% per year. There is some evidence to suggest that female condoms may provide an equivalent level of protection. Application of a vaginal gel containing tenofovir (a reverse transcriptase inhibitor) immediately before sex seems to reduce infection rates by approximately 40% among African women. By contrast, use of the spermicide nonoxynol-9 may increase the risk of transmission due to its tendency to cause vaginal and rectal irritation.

Circumcision in Sub-Saharan Africa "reduces the acquisition of HIV by heterosexual men by between 38% and 66% over 24 months". Due to these studies, both the World Health Organization and UNAIDS recommended male circumcision as a method of preventing female-to-male HIV transmission in 2007 in areas with a high rates of HIV. However, whether it protects against male-to-female transmission is disputed, and whether it is of benefit in developed countries and among men who have sex with men is undetermined. The International Antiviral Society, however, does recommend for all sexually active heterosexual males and that it be discussed as an option with men who have sex with men. Some experts fear that a lower perception of vulnerability among circumcised men may cause more sexual risk-taking behavior, thus negating its preventive effects.

Programs encouraging sexual abstinence do not appear to affect subsequent HIV risk. Evidence of any benefit from peer education is equally poor. Comprehensive sexual education provided at school may decrease high risk behavior. A substantial minority of young people continues to engage in high-risk practices despite knowing about HIV/AIDS, underestimating their own risk of becoming infected with HIV. Voluntary counseling and testing people for HIV does not affect risky behavior in those who test negative but does increase condom use in those who test positive. It is not known whether treating other sexually transmitted infections is effective in preventing HIV.

Pre-Exposure

Antiretroviral treatment among people with HIV whose CD4 count ≤ 550 cells/µL is a very effective way to prevent HIV infection of their partner (a strategy known as treatment as prevention, or TASP). TASP is associated with a 10 to 20 fold reduction in transmission risk. Pre-exposure prophylaxis (PrEP) with a daily dose of the medications tenofovir, with or without emtricitabine, is effective in a number of groups including men who have sex with men, couples where one is HIV positive, and young heterosexuals in Africa. It may also be effective in intravenous drug users with a study finding a decrease in risk of 0.7 to 0.4 per 100 person years.

Universal precautions within the health care environment are believed to be effective in decreasing the risk of HIV. Intravenous drug use is an important risk factor and harm reduction strategies such as needle-exchange programs and opioid substitution therapy appear effective in decreasing this risk.

Post-Exposure

A course of antiretrovirals administered within 48 to 72 hours after exposure to HIV-positive blood or genital secretions is referred to as post-exposure prophylaxis (PEP). The use of the single agent zidovudine reduces the risk of a HIV infection five-fold following a needle-stick injury. As of 2013, the prevention regimen recommended in the United States consists of three medications—tenofovir, emtricitabine and raltegravir—as this may reduce the risk further.

PEP treatment is recommended after a sexual assault when the perpetrator is known to be HIV positive, but is controversial when their HIV status is unknown. The duration of treatment is usually four weeks and is frequently associated with adverse effects—where zidovudine is used, about 70% of cases result in adverse effects such as nausea (24%), fatigue (22%), emotional distress (13%) and headaches (9%).

Mother-To-Child

Programs to prevent the vertical transmission of HIV (from mothers to children) can reduce rates of transmission by 92–99%. This primarily involves the use of a combination of antiviral medications during pregnancy and after birth in the infant and potentially includes bottle feeding rather than breastfeeding. If replacement feeding is acceptable, feasible, affordable, sustainable, and safe, mothers should avoid breastfeeding their infants; however exclusive breastfeeding is recommended during the first months of life if this is not the case. If exclusive breastfeeding is carried out, the provision of extended antiretroviral prophylaxis to the infant decreases the risk of transmission. In 2015, Cuba became the first country in the world to eradicate mother-to-child transmission of HIV.

Vaccination

Currently, there is no licensed vaccine for HIV or AIDS. The most effective vaccine trial to date, RV 144, was published in 2009 and found a partial reduction in the risk of transmission of roughly 30%, stimulating some hope in the research community of developing a truly effective vaccine. Further trials of the RV 144 vaccine are ongoing.

Treatment

There is currently no cure or effective HIV vaccine. Treatment consists of highly active antiretroviral therapy (HAART) which slows progression of the disease. As of 2010 more than 6.6 million people were taking them in low and middle income countries. Treatment also includes preventive and active treatment of opportunistic infections.

Antiviral Therapy

Current HAART options are combinations (or "cocktails") consisting of at least three medications belonging to at least two types, or "classes," of antiretroviral agents. Initially treatment is typically a non-nucleoside reverse transcriptase inhibitor (NNRTI) plus two nucleoside analogue reverse transcriptase inhibitors (NRTIs). Typical NRTIs include: zidovudine (AZT) or tenofovir (TDF) and lamivudine (3TC) or emtricitabine (FTC). Combinations of agents which include protease inhibitors (PI) are used if the above regimen loses effectiveness.

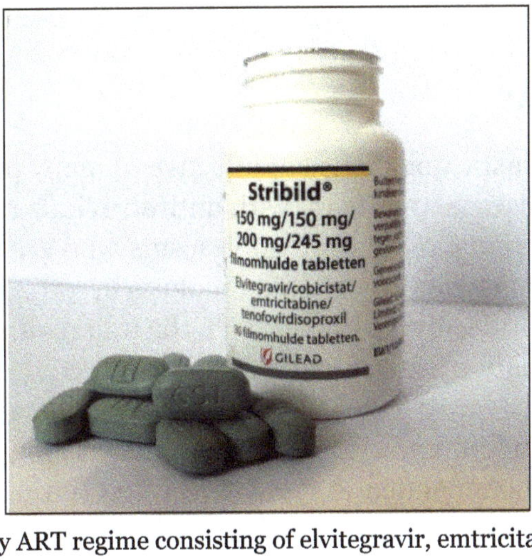

Stribild – a common once-daily ART regime consisting of elvitegravir, emtricitabine, tenofovir and the booster cobicistat.

The World Health Organization and United States recommends antiretrovirals in people of all ages including pregnant women as soon as the diagnosis is made regardless of CD4 count. Once treatment is begun it is recommended that it is continued without breaks or "holidays". Many people are diagnosed only after treatment ideally should have begun. The desired outcome of treatment is a long term plasma HIV-RNA count below 50 copies/mL. Levels to determine if treatment is effective are initially recommended after four weeks and once levels fall below 50 copies/mL checks every three to six months are typically adequate. Inadequate control is deemed to be greater than 400 copies/mL. Based on these criteria treatment is effective in more than 95% of people during the first year.

Benefits of treatment include a decreased risk of progression to AIDS and a decreased risk of death. In the developing world treatment also improves physical and mental health. With treatment there is a 70% reduced risk of acquiring tuberculosis. Additional benefits include a decreased risk of transmission of the disease to sexual partners and a decrease in mother-to-child transmission. The effectiveness of treatment depends to a large part on compliance. Reasons for non-adherence include poor access to medical care, inadequate social supports, mental illness and drug abuse. The complexity of treatment regimens (due to pill numbers and dosing frequency) and adverse effects may reduce adherence. Even though cost is an important issue with some medications, 47% of those who needed them were taking them in low and middle income countries as of 2010 and the rate of adherence is similar in low-income and high-income countries.

Specific adverse events are related to the antiretroviral agent taken. Some relatively common adverse events include: lipodystrophy syndrome, dyslipidemia, and diabetes mellitus, especially with protease inhibitors. Other common symptoms include diarrhea, and an increased risk of cardiovascular disease. Newer recommended treatments are associated with fewer adverse effects. Certain medications may be associated with birth defects and therefore may be unsuitable for women hoping to have children.

Treatment recommendations for children are somewhat different from those for adults. The World Health Organisation recommends treating all children less than 5 years of age; children above 5 are treated like adults. The United States guidelines recommend treating all children less than 12 months of age and all those with HIV RNA counts greater than 100,000 copies/mL between one year and five years of age.

Opportunistic Infections

Measures to prevent opportunistic infections are effective in many people with HIV/AIDS. In addition to improving current disease, treatment with antiretrovirals reduces the risk of developing additional opportunistic infections. Adults and adolescents who are living with HIV (even on anti-retroviral therapy) with no evidence of active tuberculosis in settings with high tuberculosis burden should receive isoniazid preventive therapy (IPT), the tuberculin skin test can be used to help decide if IPT is needed. Vaccination against hepatitis A and B is advised for all people at risk of HIV before they become infected; however it may also be given after infection. Trimethoprim/sulfamethoxazole prophylaxis between four and six weeks of age and ceasing breastfeeding in infants born to HIV positive mothers is recommended in resource limited settings. It is also recommended to prevent PCP when a person's CD4 count is below 200 cells/uL and in those who have or have previously had PCP. People with substantial immunosuppression are also advised to receive prophylactic therapy for toxoplasmosis and Cryptococcus meningitis. Appropriate preventive measures have reduced the rate of these infections by 50% between 1992 and 1997.

Diet

The World Health Organization (WHO) has issued recommendations regarding nutrient requirements in HIV/AIDS. A generally healthy diet is promoted. Some evidence has shown a benefit from micronutrient supplements. Evidence for supplementation with selenium is mixed with some tentative evidence of benefit. There is some evidence that vitamin A supplementation in children reduces mortality and improves growth. In Africa in nutritionally compromised pregnant and lactating women a multivitamin supplementation has improved outcomes for both mothers and children. Dietary intake of micronutrients at RDA levels by HIV-infected adults is recommended by the WHO; higher intake of vitamin A, zinc, and iron can produce adverse effects in HIV positive adults, and is not recommended unless there is documented deficiency.

Alternative Medicine

In the US, approximately 60% of people with HIV use various forms of complementary or alternative medicine, even though the effectiveness of most of these therapies has not been established. There is not enough evidence to support the use of herbal medicines. There is insufficient evidence to recommend or support the use of medical cannabis to try to increase appetite or weight gain.

Prognosis

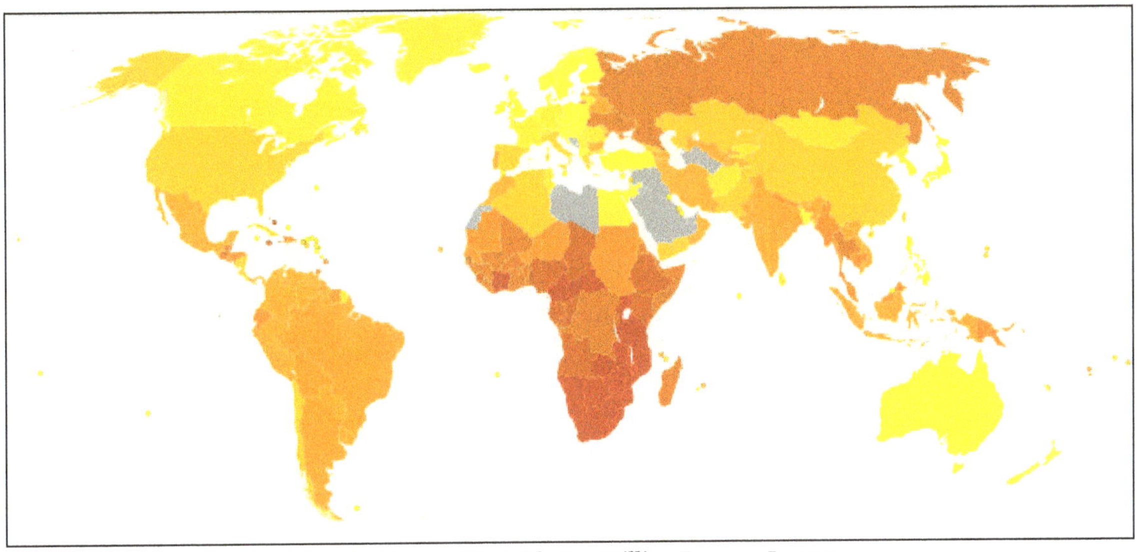

Deaths Due To Hiv/Aids Per Million Persons In 2012.

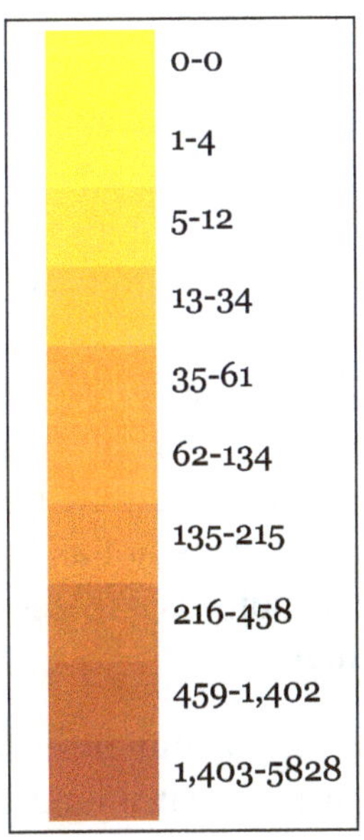

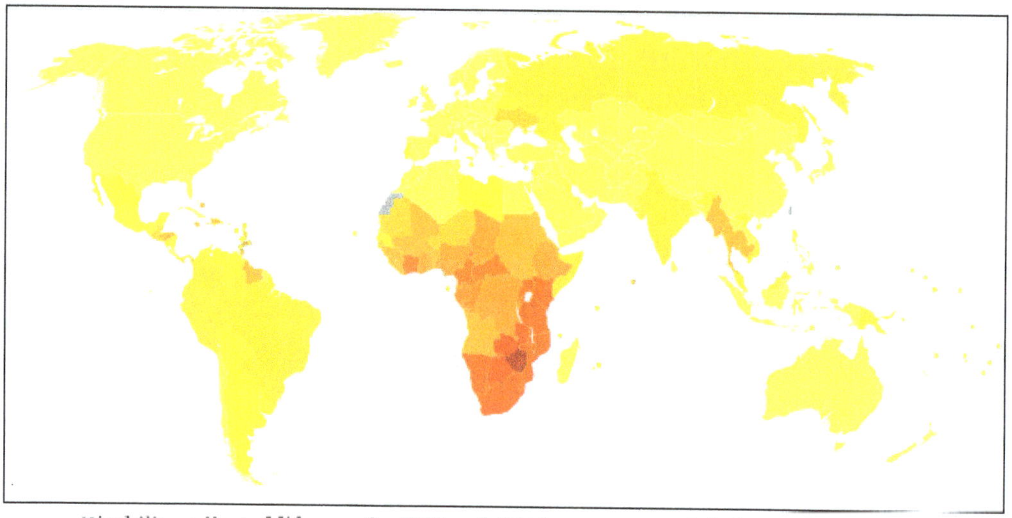

Disability-adjusted life year for HIV and AIDS per 100,000 inhabitants as of 2004.

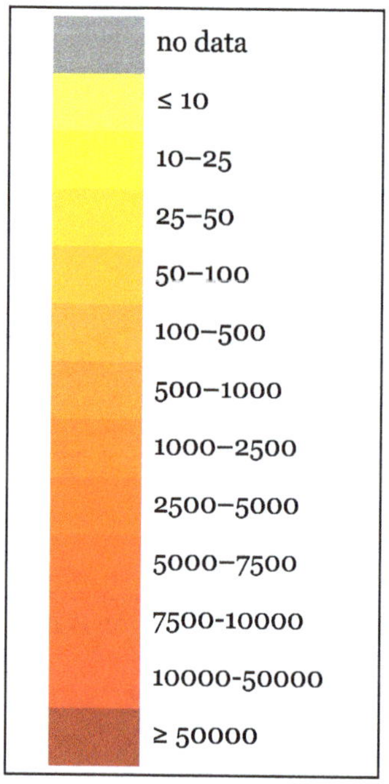

HIV/AIDS has become a chronic rather than an acutely fatal disease in many areas of the world. Prognosis varies between people, and both the CD4 count and viral load are useful for predicted outcomes. Without treatment, average survival time after infection with HIV is estimated to be 9 to 11 years, depending on the HIV subtype. After the diagnosis of AIDS, if treatment is not available, survival ranges between 6 and 19 months. HAART and appropriate prevention of opportunistic infections reduces the death rate by 80%, and raises the life expectancy for a newly diagnosed young adult to 20–50 years. This is between two thirds and nearly that of the general population. If treatment is started late in the infection, prognosis is not as good: for example, if treatment is begun following the diagnosis of AIDS, life expectancy is ~10–40 years. Half of infants born with HIV die before two years of age without treatment.

The primary causes of death from HIV/AIDS are opportunistic infections and cancer, both of which are frequently the result of the progressive failure of the immune system. Risk of cancer appears to increase once the CD4 count is below 500/μL. The rate of clinical disease progression varies widely between individuals and has been shown to be affected by a number of factors such as a person's susceptibility and immune function; their access to health care, the presence of co-infections; and the particular strain (or strains) of the virus involved.

Tuberculosis co-infection is one of the leading causes of sickness and death in those with HIV/AIDS being present in a third of all HIV-infected people and causing 25% of HIV-related deaths. HIV is also one of the most important risk factors for tuberculosis. Hepatitis C is another very common co-infection where each disease increases the progression of the other. The two most common cancers associated with HIV/AIDS are Kaposi's sarcoma and AIDS-related non-Hodgkin's lymphoma.

Even with anti-retroviral treatment, over the long term HIV-infected people may experience neurocognitive disorders, osteoporosis, neuropathy, cancers, nephropathy, and cardiovascular disease. Some conditions like lipodystrophy may be caused both by HIV and its treatment.

Epidemiology

HIV/AIDS is a global pandemic. As of 2014, approximately 37 million people have HIV worldwide with the number of new infections that year being about 2 million. This is down from 3.1 million new infections in 2001. Of these 37 million more than half are women and 2.6 million are less than 15 years old. It resulted in about 1.2 million deaths in 2014, down from a peak of 2.2 million in 2005.

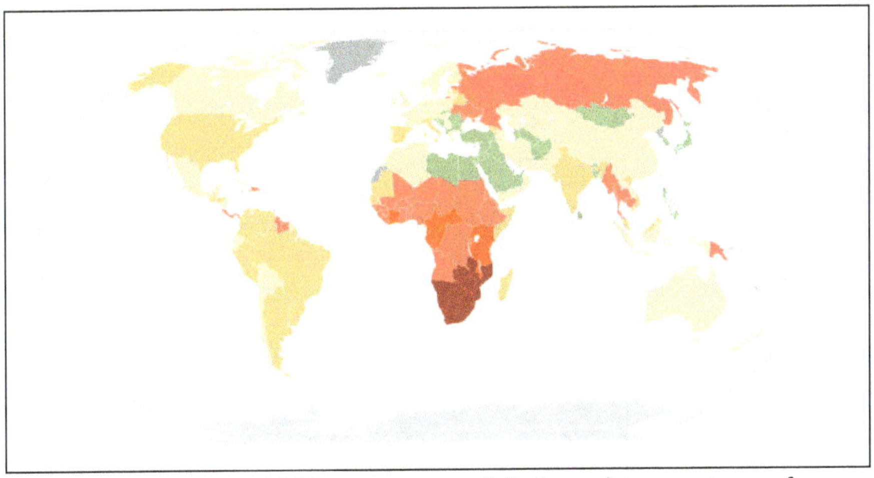

Estimated percentage of HIV among young adults (15–49) per country as of 2011.

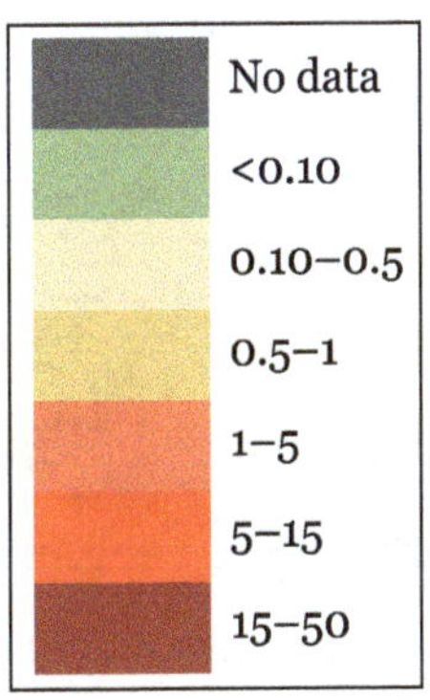

Sub-Saharan Africa is the region most affected. In 2010, an estimated 68% (22.9 million) of all HIV cases and 66% of all deaths (1.2 million) occurred in this region. This means that about 5% of the adult population is infected and it is believed to be the cause of 10% of all deaths in children. Here in contrast to other regions women compose nearly 60% of cases. South Africa has the largest population of people with HIV of any country in the world at 5.9 million. Life expectancy has fallen in the worst-affected countries due to HIV/AIDS; for example, in 2006 it was estimated that it had dropped from 65 to 35 years in Botswana. Mother-to-child transmission, as of 2013, in Botswana and South Africa has decreased to less than 5% with improvement in many other African nations due to improved access to antiretroviral therapy.

South & South East Asia is the second most affected; in 2010 this region contained an estimated 4 million cases or 12% of all people living with HIV resulting in approximately 250,000 deaths. Approximately 2.4 million of these cases are in India.

In 2008 in the United States approximately 1.2 million people were living with HIV, resulting in about 17,500 deaths. The US Centers for Disease Control and Prevention estimated that in 2008 20% of infected Americans were unaware of their infection. In the United Kingdom as of 2009 there were approximately 86,500 cases which resulted in 516 deaths. In Canada as of 2008 there were about 65,000 cases causing 53 deaths. Between the first recognition of AIDS in 1981 and 2009 it has led to nearly 30 million deaths. Prevalence is lowest in Middle East and North Africa at 0.1% or less, East Asia at 0.1% and Western and Central Europe at 0.2%. The worst affected European countries, in 2009 and 2012 estimates, are Russia, Ukraine, Latvia, Moldova, Portugal and Belarus, in decreasing order of prevalence.

History

Discovery

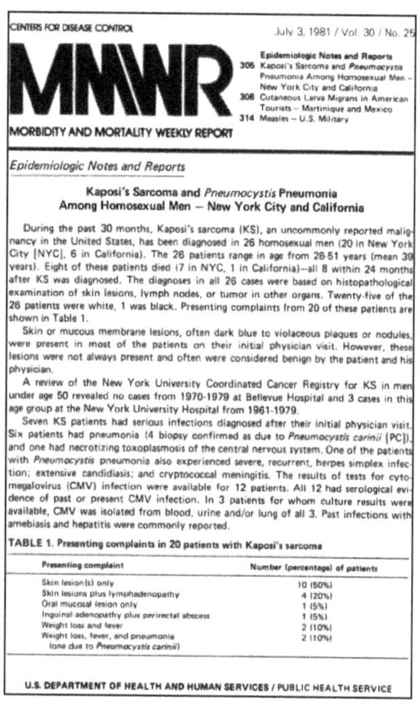

CENTERS FOR DISEASE CONTROL

July 3, 1981 / Vol. 30 / No. 25

MMWR

MORBIDITY AND MORTALITY WEEKLY REPORT

Epidemiologic Notes and Reports

305 Kaposi's Sarcoma and *Pneumocystis* Pneumonia Among Homosexual Men – New York City and California
308 Cutaneous Larva Migrans in American Tourists – Martinique and Mexico
314 Measles – U.S. Military

Epidemiologic Notes and Reports

Kaposi's Sarcoma and *Pneumocystis* Pneumonia Among Homosexual Men – New York City and California

During the past 30 months, Kaposi's sarcoma (KS), an uncommonly reported malignancy in the United States, has been diagnosed in 26 homosexual men (20 in New York City [NYC], 6 in California). The 26 patients range in age from 26-51 years (mean 39 years). Eight of these patients died (7 in NYC, 1 in California)—all 8 within 24 months after KS was diagnosed. The diagnoses in all 26 cases were based on histopathological examination of skin lesions, lymph nodes, or tumor in other organs. Twenty-five of the 26 patients were white, 1 was black. Presenting complaints from 20 of these patients are shown in Table 1.

Skin or mucous membrane lesions, often dark blue to violaceous plaques or nodules, were present in most of the patients on their initial physician visit. However, these lesions were not always present and often were considered benign by the patient and his physician.

A review of the New York University Coordinated Cancer Registry for KS in men under age 50 revealed no cases from 1970-1979 at Bellevue Hospital and 3 cases in this age group at the New York University Hospital from 1961-1979.

Seven KS patients had serious infections diagnosed after their initial physician visit. Six patients had pneumonia (4 biopsy confirmed as due to *Pneumocystis carinii* [PC]), and one had necrotizing toxoplasmosis of the central nervous system. One of the patients with *Pneumocystis* pneumonia also experienced severe, recurrent, herpes simplex infection; extensive candidiasis; and cryptococcal meningitis. The results of tests for cytomegalovirus (CMV) infection were available for 12 patients. All 12 had serological evidence of past or present CMV infection. In 3 patients for whom culture results were available, CMV was isolated from blood, urine and/or lung of all 3. Past infections with amebiasis and hepatitis were commonly reported.

TABLE 1. Presenting complaints in 20 patients with Kaposi's sarcoma

Presenting complaint	Number (percentage) of patients
Skin lesion(s) only	10 (50%)
Skin lesions plus lymphadenopathy	4 (20%)
Oral mucosal lesion only	1 (5%)
Inguinal adenopathy plus perirectal abscess	1 (5%)
Weight loss and fever	2 (10%)
Weight loss, fever, and pneumonia (one due to *Pneumocystis carinii*)	2 (10%)

U.S. DEPARTMENT OF HEALTH AND HUMAN SERVICES / PUBLIC HEALTH SERVICE

The *Morbidity and Mortality Weekly Report* reported in 1981 on what was later to be called "AIDS".

AIDS was first clinically observed in 1981 in the United States. The initial cases were a cluster of injecting drug users and homosexual men with no known cause of impaired immunity who showed symptoms of *Pneumocystis carinii* pneumonia (PCP), a rare opportunistic infection that was known to occur in people with very compromised immune systems. Soon thereafter, an unexpected number of homosexual men developed a previously rare skin cancer called Kaposi's sarcoma (KS). Many more cases of PCP and KS emerged, alerting U.S. Centers for Disease Control and Prevention (CDC) and a CDC task force was formed to monitor the outbreak.

In the early days, the CDC did not have an official name for the disease, often referring to it by way of the diseases that were associated with it, for example, lymphadenopathy, the disease after which the discoverers of HIV originally named the virus. They also used *Kaposi's sarcoma and opportunistic infections*, the name by which a task force had been set up in 1981. At one point, the CDC coined the phrase "the 4H disease", since the syndrome seemed to affect heroin users, homosexuals, hemophiliacs, and Haitians. In the general press, the term "GRID", which stood for gay-related immune deficiency, had been coined. However, after determining that AIDS was not isolated to the gay community, it was realized that the term GRID was misleading and the term AIDS was introduced at a meeting in July 1982. By September 1982 the CDC started referring to the disease as AIDS.

In 1983, two separate research groups led by Robert Gallo and Luc Montagnier declared that a novel retrovirus may have been infecting people with AIDS, and published their findings in the same issue of the journal *Science*. Gallo claimed that a virus his group had isolated from a person with AIDS was strikingly similar in shape to other human T-lymphotropic viruses (HTLVs) his group had been the first to isolate. Gallo's group called their newly isolated virus HTLV-III. At the same time, Montagnier's group isolated a virus from a person presenting with swelling of the lymph nodes of the neck and physical weakness, two characteristic symptoms of AIDS. Contradicting the report from Gallo's group, Montagnier and his colleagues showed that core proteins of this virus were immunologically different from those of HTLV-I. Montagnier's group named their isolated virus lymphadenopathy-associated virus (LAV). As these two viruses turned out to be the same, in 1986, LAV and HTLV-III were renamed HIV.

Origins

Both HIV-1 and HIV-2 are believed to have originated in non-human primates in West-central Africa and were transferred to humans in the early 20th century. HIV-1 appears to have originated in southern Cameroon through the evolution of SIV(cpz), a simian immunodeficiency virus (SIV) that infects wild chimpanzees (HIV-1 descends from the SIVcpz endemic in the chimpanzee subspecies *Pan troglodytes troglodytes*). The closest relative of HIV-2 is SIV(smm), a virus of the sooty mangabey (*Cercocebus atys atys*), an Old World monkey living in coastal West Africa (from southern Senegal to western Côte d'Ivoire). New World monkeys such as the owl monkey are resistant to HIV-1 infection, possibly because of a genomic fusion of two viral resistance genes. HIV-1 is thought to have jumped the species barrier on at least three separate occasions, giving rise to the three groups of the virus, M, N, and O.

There is evidence that humans who participate in bushmeat activities, either as hunters or as bushmeat vendors, commonly acquire SIV. However, SIV is a weak virus which is typically suppressed by the human immune system within weeks of infection. It is thought that several transmissions of the virus from individual to individual in quick succession are necessary to allow it enough time to mutate into HIV. Furthermore, due to its relatively low person-to-person transmission rate, SIV can only spread throughout the population in the presence of one or more high-risk transmission channels, which are thought to have been absent in Africa before the 20th century.

Left to right: the African green monkey source of SIV, the sooty mangabey source of HIV-2 and the chimpanzee source of HIV-1.

Specific proposed high-risk transmission channels, allowing the virus to adapt to humans and spread throughout the society, depend on the proposed timing of the animal-to-human crossing. Genetic studies of the virus suggest that the most recent common ancestor of the HIV-1 M group dates back to circa 1910. Proponents of this dating link the HIV epidemic with the emergence of colonialism and growth of large colonial African cities, leading to social changes, including a higher degree of sexual promiscuity, the spread of prostitution, and the accompanying high frequency of genital ulcer diseases (such as syphilis) in nascent colonial cities. While transmission rates of HIV during vaginal intercourse are low under regular circumstances, they are increased many fold if one of the partners suffers from a sexually transmitted infection causing genital ulcers. Early 1900s colonial cities were notable due to their high prevalence of prostitution and genital ulcers, to the degree that, as of 1928, as many as 45% of female residents of eastern Kinshasa were thought to have been prostitutes, and, as of 1933, around 15% of all residents of the same city had syphilis.

An alternative view holds that unsafe medical practices in Africa after World War II, such as unsterile reuse of single use syringes during mass vaccination, antibiotic and anti-malaria treatment campaigns, were the initial vector that allowed the virus to adapt to humans and spread.

The earliest well-documented case of HIV in a human dates back to 1959 in the Congo. The earliest retrospectively described case of AIDS is believed to have been in Norway beginning in 1966. In July 1960, in the wake its independence, the United Nations recruited Francophone experts and technicians from all over the world to assist in filling administrative gaps left by Belgium, who did not leave behind an African elite to run the country. By 1962, Haitians made up the second largest group of well-educated experts (out of the 48 national groups recruited), that totaled around 4500 in the country. Dr. Jacques Pépin, a Quebecer author of *The Origins of AIDS*, stipulates that Haiti was one of HIV's entry points to the United States and that one of them may have carried HIV back across the Atlantic in the 1960s. Although, the virus may have been present in the United States as early as 1966, the vast majority of infections occurring outside sub-Saharan Africa (including the U.S.) can be traced back to a single unknown individual who became infected with HIV in Haiti and then brought the infection to the United States some time around 1969. The epidemic then rapidly spread among high-risk groups (initially, sexually promiscuous men who have sex with men). By 1978, the prevalence of HIV-1 among homosexual male residents of New York and San Francisco was estimated at 5%, suggesting that several thousand individuals in the country had been infected.

Society and Culture

Stigma

AIDS stigma exists around the world in a variety of ways, including ostracism, rejection, discrimination and avoidance of HIV infected people; compulsory HIV testing without prior consent or protection of confidentiality; violence against HIV infected individuals or people who are perceived to be infected with HIV; and the quarantine of HIV infected individuals. Stigma-related violence or the fear of violence prevents many people from seeking HIV testing, returning for their results, or securing treatment, possibly turning what could be a manageable chronic illness into a death sentence and perpetuating the spread of HIV.

Ryan White became a poster child for HIV after being expelled from school because he was infected.

AIDS stigma has been further divided into the following three categories:

- *Instrumental AIDS stigma*—a reflection of the fear and apprehension that are likely to be associated with any deadly and transmissible illness.
- *Symbolic AIDS stigma*—the use of HIV/AIDS to express attitudes toward the social groups or lifestyles perceived to be associated with the disease.
- *Courtesy AIDS stigma*—stigmatization of people connected to the issue of HIV/AIDS or HIV-positive people.

Often, AIDS stigma is expressed in conjunction with one or more other stigmas, particularly those associated with homosexuality, bisexuality, promiscuity, prostitution, and intravenous drug use.

In many developed countries, there is an association between AIDS and homosexuality or bisexuality, and this association is correlated with higher levels of sexual prejudice, such as anti-homosexual/bisexual attitudes. There is also a perceived association between AIDS and all male-male sexual behavior, including sex between uninfected men. However, the dominant mode of spread worldwide for HIV remains heterosexual transmission.

In 2003, as part of an overall reform of marriage and population legislation, it became legal for people with AIDS to marry in China.

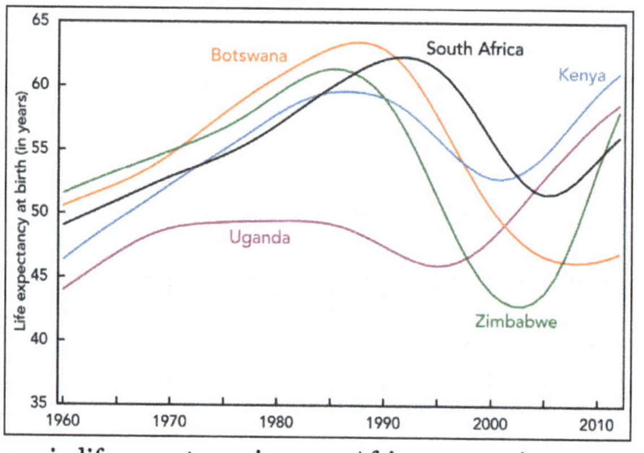

Changes in life expectancy in some African countries, 1960–2012.

HIV/AIDS affects the economics of both individuals and countries. The gross domestic product of the most affected countries has decreased due to the lack of human capital. Without proper nutrition, health care and medicine, large numbers of people die from AIDS-related complications. They will not only be unable to work, but will also require significant medical care. It is estimated that as of 2007 there were 12 million AIDS orphans. Many are cared for by elderly grandparents.

Returning to work after beginning treatment for HIV/AIDS is difficult, and affected people often work less than the average worker. Unemployment in people with HIV/AIDS also is associated with suicidal ideation, memory problems, and social isolation; employment increases self-esteem, sense of dignity, confidence, and quality of life. A 2015 Cochrane review found low-quality evidence that antiretroviral treatment helps people with HIV/AIDS work more, and increases the chance that a person with HIV/AIDS will be employed.

By affecting mainly young adults, AIDS reduces the taxable population, in turn reducing the resources available for public expenditures such as education and health services not related to AIDS resulting in increasing pressure for the state's finances and slower growth of the economy. This causes a slower growth of the tax base, an effect that is reinforced if there are growing expenditures on treating the sick, training (to replace sick workers), sick pay and caring for AIDS orphans. This is especially true if the sharp increase in adult mortality shifts the responsibility and blame from the family to the government in caring for these orphans.

At the household level, AIDS causes both loss of income and increased spending on healthcare. A study in Côte d'Ivoire showed that households having a person with HIV/AIDS spent twice as much on medical expenses as other households. This additional expenditure also leaves less income to spend on education and other personal or family investment.

Religion and AIDS

The topic of religion and AIDS has become highly controversial in the past twenty years, primarily because some religious authorities have publicly declared their opposition to the use of condoms.

The religious approach to prevent the spread of AIDS according to a report by American health expert Matthew Hanley titled *The Catholic Church and the Global AIDS Crisis* argues that cultural changes are needed including a re-emphasis on fidelity within marriage and sexual abstinence outside of it.

Some religious organisations have claimed that prayer can cure HIV/AIDS. In 2011, the BBC reported that some churches in London were claiming that prayer would cure AIDS, and the Hackney-based Centre for the Study of Sexual Health and HIV reported that several people stopped taking their medication, sometimes on the direct advice of their pastor, leading to a number of deaths. The Synagogue Church Of All Nations advertise an "anointing water" to promote God's healing, although the group deny advising people to stop taking medication.

Media Portrayal

One of the first high-profile cases of AIDS was the American Rock Hudson, a gay actor who had been married and divorced earlier in life, who died on October 2, 1985 having announced that he was suffering from the virus on July 25 that year. He had been diagnosed during 1984. A notable British casualty of AIDS that year was Nicholas Eden, a gay politician and son of the late prime minister Anthony Eden. On November 24, 1991, the virus claimed the life of British rock star Freddie Mercury, lead singer of the band Queen, who died from an AIDS-related illness having only revealed the diagnosis on the previous day. However, he had been diagnosed as HIV positive in 1987. One of the first high-profile heterosexual cases of the virus was Arthur Ashe, the American tennis player. He was diagnosed as HIV positive on August 31, 1988, having contracted the virus from blood transfusions during heart surgery earlier in the 1980s. Further tests within 24 hours of the initial diagnosis revealed that Ashe had AIDS, but he did not tell the public about his diagnosis until April 1992. He died as a result on February 6, 1993 at age 49.

Therese Frare's photograph of gay activist David Kirby, as he lay dying from AIDS while surrounded by family, was taken in April 1990. *LIFE magazine* said the photo became the one image "most powerfully identified with the HIV/AIDS epidemic." The photo was displayed in *LIFE magazine*, was the winner of the World Press Photo, and acquired worldwide notoriety after being used in a United Colors of Benetton advertising campaign in 1992. In 1996, Johnson Aziga, a Ugandan-born Canadian was diagnosed with HIV, but subsequently had unprotected sex with 11 women without disclosing his diagnosis. By 2003 seven had contracted HIV, and two died from complications related to AIDS. Aziga was convicted of first-degree murder and is liable to a life sentence.

Criminal Transmission

Criminal transmission of HIV is the intentional or reckless infection of a person with the human immunodeficiency virus (HIV). Some countries or jurisdictions, including some areas of the United States, have laws that criminalize HIV transmission or exposure. Others may charge the accused under laws enacted before the HIV pandemic.

Misconceptions

There are many misconceptions about HIV and AIDS. Three of the most common are that AIDS can spread through casual contact, that sexual intercourse with a virgin will cure AIDS, and that

HIV can infect only gay men and drug users. In 2014, some among the British public wrongly thought you could get HIV from kissing (16%), sharing a glass (5%), spitting (16%), a public toilet seat (4%), and coughing or sneezing (5%). Other misconceptions are that any act of anal intercourse between two uninfected gay men can lead to HIV infection, and that open discussion of HIV and homosexuality in schools will lead to increased rates of AIDS.

A small group of individuals continue to dispute the connection between HIV and AIDS, the existence of HIV itself, or the validity of HIV testing and treatment methods. These claims, known as AIDS denialism, have been examined and rejected by the scientific community. However, they have had a significant political impact, particularly in South Africa, where the government's official embrace of AIDS denialism (1999–2005) was responsible for its ineffective response to that country's AIDS epidemic, and has been blamed for hundreds of thousands of avoidable deaths and HIV infections.

Several discredited conspiracy theories have held that HIV was created by scientists, either inadvertently or deliberately. Operation INFEKTION was a worldwide Soviet active measures operation to spread the claim that the United States had created HIV/AIDS. Surveys show that a significant number of people believed – and continue to believe – in such claims.

Research

HIV/AIDS research includes all medical research which attempts to prevent, treat, or cure HIV/AIDS along with fundamental research about the nature of HIV as an infectious agent and AIDS as the disease caused by HIV.

Many governments and research institutions participate in HIV/AIDS research. This research includes behavioral health interventions such as sex education, and drug development, such as research into microbicides for sexually transmitted diseases, HIV vaccines, and antiretroviral drugs. Other medical research areas include the topics of pre-exposure prophylaxis, post-exposure prophylaxis, and circumcision and HIV.

Chronic Granulomatous Disease

Chronic granulomatous disease (CGD) (also known as Bridges–Good syndrome, Chronic granulomatous disorder, and Quie syndrome) is a diverse group of hereditary diseases in which certain cells of the immune system have difficulty forming the reactive oxygen compounds (most importantly the superoxide radical due to defective phagocyte NADPH oxidase) used to kill certain ingested pathogens. This leads to the formation of granulomata in many organs. CGD affects about 1 in 200,000 people in the United States, with about 20 new cases diagnosed each year.

This condition was first discovered in 1950 in a series of 4 boys from Minnesota, and in 1957 was named "a fatal granulomatosus of childhood" in a publication describing their disease. The underlying cellular mechanism that causes chronic granulomatous disease was discovered in 1967, and research since that time has further elucidated the molecular mechanisms underlying the disease. Bernard Babior made key contributions in linking the defect of superoxide production of white blood cells, to the etiology of the disease. In 1986, the X-linked form of CGD was the first disease for which positional cloning was used to identify the underlying genetic mutation.

Classification

Chronic granulomatous disease is the name for a genetically heterogeneous group of immunodeficiencies. The core defect is a failure of phagocytic cells to kill organisms that they have engulfed because of defects in a system of enzymes that produce free radicals and other toxic small molecules. There are several types, including:

- X-linked chronic granulomatous disease]] (CGD).
- Autosomal recessive cytochrome b-negative CGD.
- Autosomal recessive cytochrome b-positive CGD type I.
- Autosomal recessive cytochrome b-positive CGD type II.
- Atypical granulomatous disease.

Symptoms

Classically, patients with chronic granulomatous disease will suffer from recurrent bouts of infection due to the decreased capacity of their immune system to fight off disease-causing organisms. The recurrent infections they acquire are specific and are, in decreasing order of frequency:

- Pneumonia.
- Abscesses of the skin, tissues, and organs.
- Suppurative arthritis.
- Osteomyelitis.
- Bacteremia/fungemia.
- Superficial skin infections such as cellulitis or impetigo.

Most people with CGD are diagnosed in childhood, usually before age 5. Early diagnosis is important since these people can be placed on antibiotics to ward off infections before they occur. Small groups of CGD patients may also be affected by McLeod syndrome because of the proximity of the two genes on the same X-chromosome.

Atypical Infections

People with CGD are sometimes infected with organisms that usually do not cause disease in people with normal immune systems. Among the most common organisms that cause disease in CGD patients are.

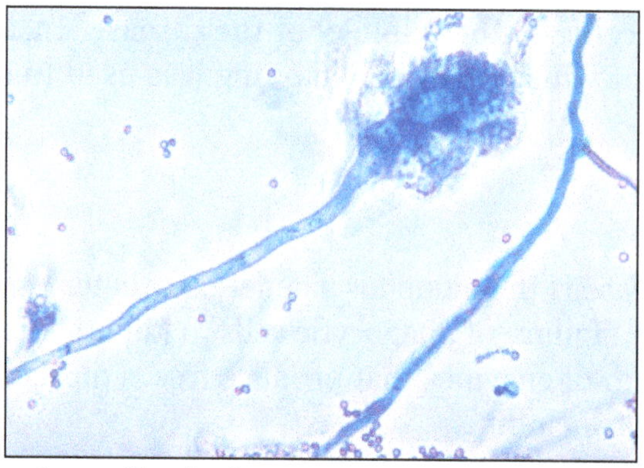

Microscopic image of the fungus, *Aspergillus fumigatus*, an organism that commonly causes disease in people with chronic granulomatous disease.

- Bacteria (particularly those that are catalase-positive):
    - *Staphylococcus aureus.*
    - *Serratia marcescens.*
    - *Listeria* species.
    - *E. coli.*
    - *Klebsiella* species.
    - *Pseudomonas cepacia, a.k.a. Burkholderia cepacia.*
    - *Nocardia.*
- Fungi:
    - *Aspergillus* species. Aspergillus has a propensity to cause infection in people with CGD and of the Aspergillus species, *Aspergillus fumigatus* seems to be most common in CGD.
    - *Candida* species.

Patients with CGD can usually resist infections of catalase-negative bacteria. Catalase is an enzyme that catalyzes the breakdown of hydrogen peroxide in many organisms. In organisms that lack catalase (catalase-negative), normal metabolic functions will cause an accumulation of hydrogen peroxide which the host's immune system can use to fight off the infection. In organisms that have catalase (catalase-positive), the enzyme breaks down any hydrogen peroxide that was produced through normal metabolism. Therefore, hydrogen peroxide will not accumulate, leaving the patient vulnerable to catalase-positive bacteria.

Genetics

Most cases of chronic granulomatous disease are transmitted as a mutation on the X chromosome and are thus called an "X-linked trait". The affected gene on the X chromosome codes for the gp91 protein p91-PHOX (*p* is the weight of the protein in kDa; the *g* means glycoprotein). CGD can also be transmitted in an autosomal recessive fashion (via CYBA and NCF1) and affects other PHOX proteins. The type of mutation that causes both types of CGD are varied and may be deletions, frame-shift, nonsense, and missense.

A low level of NADPH, the cofactor required for superoxide synthesis, can lead to CGD. This has been reported in women who are homozygous for the genetic defect causing glucose-6-phosphate dehydrogenase deficiency (G6PD), which is characterised by reduced NADPH levels.

Pathophysiology

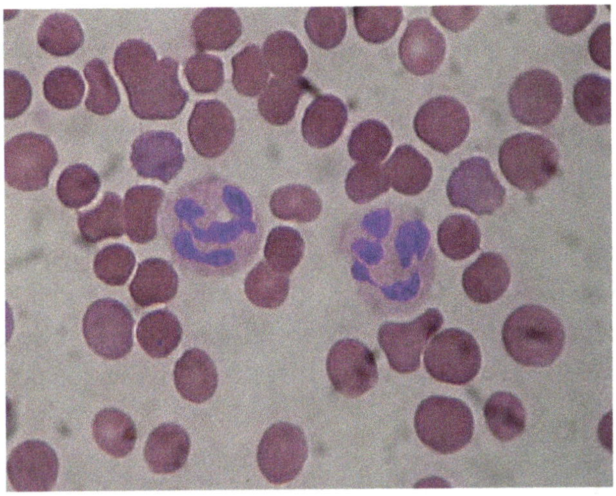

Two neutrophils among many red blood cells. Neutrophils are one type of cell affected by chronic granulomatous disease.

Phagocytes (i.e., neutrophils and macrophages) require an enzyme to produce reactive oxygen species to destroy bacteria after they are ingested (phagocytosis), a process known as the respiratory burst. This enzyme is termed "phagocyte NADPH oxidase" (*PHOX*). This enzyme oxidizes NADPH and reduces molecular oxygen to produce superoxide anions, a reactive oxygen species. Superoxide is then disproportionated into peroxide and molecular oxygen by superoxide dismutase. Finally, peroxide is used by myeloperoxidase to oxidize chloride ions into hypochlorite (the active component of bleach), which is toxic to bacteria. Thus, NADPH oxidase is critical for phagocyte killing of bacteria through reactive oxygen species.

(Two other mechanisms are used by phagocytes to kill bacteria: nitric oxide and proteases, but the loss of ROS-mediated killing alone is sufficient to cause chronic granulomatous disease.)

Defects in one of the four essential subunits of phagocyte NADPH oxidase (PHOX) can all cause CGD of varying severity, dependent on the defect. There are over 410 known possible defects in the PHOX enzyme complex that can lead to chronic granulomatous disease.

Diagnosis

The nitroblue-tetrazolium (NBT) test is the original and most widely known test for chronic granulomatous disease. It is negative in CGD, meaning that it does not turn blue. The higher the blue score, the better the cell is at producing reactive oxygen species. This test depends upon the direct reduction of NBT to the insoluble blue compound formazan by NADPH oxidase; NADPH is oxidized in the same reaction. This test is simple to perform and gives rapid results, but only tells whether or not there is a problem with the PHOX enzymes, not how much they are affected. A similar test uses dihydrorhodamine (DHR) where whole blood is stained with DHR, incubated, and stimulated to produce superoxide radicals which oxidize DHR to rhodamin in cells with normal function. An advanced test called the cytochrome C reduction assay tells physicians how much superoxide a patient's phagocytes can produce. Once the diagnosis of CGD is established, a genetic analysis may be used to determine exactly which mutation is the underlying cause.

Treatment

Management of chronic granulomatous disease revolves around two goals: 1) diagnose the disease early so that antibiotic prophylaxis can be given to keep an infection from occurring, and 2) educate the patient about his or her condition so that prompt treatment can be given if an infection occurs.

Antibiotics

Physicians often prescribe the antibiotic trimethoprim-sulfamethoxazole to prevent bacterial infections. This drug also has the benefit of sparing the normal bacteria of the digestive tract. Fungal infection is commonly prevented with itraconazole, although a newer drug of the same type called voriconazole may be more effective. The use of this drug for this purpose is still under scientific investigation.

Immunomodulation

Interferon, in the form of interferon gamma-1b (Actimmune) is approved by the Food and Drug Administration for the prevention of infection in CGD. It has been shown to reduce infections in CGD patients by 70% and to decrease their severity. Although its exact mechanism is still not entirely understood, it has the ability to give CGD patients more immune function and therefore, greater ability to fight off infections. This therapy has been standard treatment for CGD for several years.

Hematopoietic Stem Cell Transplantation (HSCT)

Hematopoietic stem cell transplantation from a matched donor is curative although not without significant risk.

Prognosis

There are currently no studies detailing the long term outcome of chronic granulomatous disease with modern treatment. Without treatment, children often die in the first decade of life. The increased severity of X-linked CGD results in a decreased survival rate of patients, as 20% of X-linked patients die of CGD-related causes by the age of 10, whereas 20% of autosomal recessive patients die by the age of 35. Recent experience from centers specializing in the care of patients with CGD suggests that the current mortality has fallen to under 3% and 1% respectively. CGD was initially termed "fatal granulomatous disease of childhood" because patients rarely survived past their first decade in the time before routine use of prophylactic antimicrobial agents. The average patient now survives at least 40 years.

Epidemiology

CGD affects about 1 in 200,000 people in the United States, with about 20 new cases diagnosed each year.

Chronic granulomatous disease affects all people of all races, however, there is limited information on prevalence outside of the United States. One survey in Sweden reported an incidence of 1 in 220,000 people, while a larger review of studies in Europe suggested a lower rate: 1 in 250,000 people.

History

This condition was first described in 1954 by Janeway, who reported five cases of the disease in children. In 1957 it was further characterized as "a fatal granulomatosus of childhood". The underlying cellular mechanism that causes chronic granulomatous disease was discovered in 1967, and research since that time has further elucidated the molecular mechanisms underlying the disease. Use of antibiotic prophylaxis, surgical abscess drainage, and vaccination led to the term "fatal" being dropped from the name of the disease as children survived into adulthood.

Research

Gene therapy is currently being studied as a possible treatment for chronic granulomatous disease. CGD is well-suited for gene therapy since it is caused by a mutation in single gene which only affects one body system (the hematopoietic system). Viruses have been used to deliver a normal gp91 gene to rats with a mutation in this gene, and subsequently the phagocytes in these rats were able to produce oxygen radicals.

In 2006, two human patients with X-linked chronic granulomatous disease underwent gene therapy and blood cell precursor stem cell transplantation to their bone marrow. Both patients recovered from their CGD, clearing pre-existing infections and demonstrating increased oxidase activity in their neutrophils. However, long-term complications and efficacy of this therapy were unknown.

In 2012, a 16-year-old boy with CGD was treated at the Great Ormond Street Hospital, London with an experimental gene therapy which temporarily reversed the CGD and allowed him to overcome a life-threatening lung disease.

References

- Kerrigan, Deanna (2012). The Global HIV Epidemics among Sex Workers. World Bank Publications. p. 1. ISBN 978-0-8213-9775-6
- Stürchler, Dieter A. (2006). Exposure a guide to sources of infections. Washington, DC: ASM Press. p. 544. ISBN 978-1-55581-376-5
- Martínez, edited by Miguel Angel (2010). RNA interference and viruses : current innovations and future trends. Norfolk: Caister Academic Press. p. 73. ISBN 978-1-904455-56-1
- Guideline on when to start antiretroviral therapy and on pre-exposure prophylaxis for HIV. (PDF). WHO. 2015. p. 13. ISBN 9789241509565
- Aral, Sevgi (2013). The New Public Health and STD/HIV Prevention: Personal, Public and Health Systems Approaches. Springer. p. 120. ISBN 978-1-4614-4526-5

- Harrison's Principles of Internal Medicine: Volumes 1 and 2, 18th Edition (18 ed.). McGraw-Hill Professional. 2011-08-11. ISBN 9780071748896

- MacKay, edited by Noel Richard Rose, Ian R. (2014). The autoimmune diseases (Fifth ed.). [S.l.]: Academic Press. p. Chapter 75. ISBN 978-0-12-384929-8

- Kumar, Vinay; Abbas, Abul K.; Aster, Jon C., eds. (2014). "Hypersensitivity: Immunologicaly Mediated Tissue Injury". Robbins & Cotran Pathologic Basis of Disease (9th ed.). Elsevier Health Sciences. pp. 200–11. ISBN 978-0-323-29635-9

- Cunha, Burke (2012). Antibiotic Essentials 2012 (11 ed.). Jones & Bartlett Publishers. p. 303. ISBN 9781449693831

- Pillay, Deenan; Genetti, Anna Maria; Weiss, Robin A. (2007). "Human Immunodeficiency Viruses". In Zuckerman, Arie J.; et al. Principles and practice of clinical virology (6th ed.). Hoboken, N.J.: Wiley. p. 905. ISBN 978-0-470-51799-4

- Mitchell, Richard Sheppard; Kumar, Vinay; Abbas, Abul K.; Fausto, Nelson (2007). "Table 5-1". Robbins Basic Pathology (8th ed.). Philadelphia: Saunders. ISBN 1-4160-2973-7

Permissions

All chapters in this book are published with permission under the Creative Commons Attribution Share Alike License or equivalent. Every chapter published in this book has been scrutinized by our experts. Their significance has been extensively debated. The topics covered herein carry significant information for a comprehensive understanding. They may even be implemented as practical applications or may be referred to as a beginning point for further studies.

We would like to thank the editorial team for lending their expertise to make the book truly unique. They have played a crucial role in the development of this book. Without their invaluable contributions this book wouldn't have been possible. They have made vital efforts to compile up to date information on the varied aspects of this subject to make this book a valuable addition to the collection of many professionals and students.

This book was conceptualized with the vision of imparting up-to-date and integrated information in this field. To ensure the same, a matchless editorial board was set up. Every individual on the board went through rigorous rounds of assessment to prove their worth. After which they invested a large part of their time researching and compiling the most relevant data for our readers.

The editorial board has been involved in producing this book since its inception. They have spent rigorous hours researching and exploring the diverse topics which have resulted in the successful publishing of this book. They have passed on their knowledge of decades through this book. To expedite this challenging task, the publisher supported the team at every step. A small team of assistant editors was also appointed to further simplify the editing procedure and attain best results for the readers.

Apart from the editorial board, the designing team has also invested a significant amount of their time in understanding the subject and creating the most relevant covers. They scrutinized every image to scout for the most suitable representation of the subject and create an appropriate cover for the book.

The publishing team has been an ardent support to the editorial, designing and production team. Their endless efforts to recruit the best for this project, has resulted in the accomplishment of this book. They are a veteran in the field of academics and their pool of knowledge is as vast as their experience in printing. Their expertise and guidance has proved useful at every step. Their uncompromising quality standards have made this book an exceptional effort. Their encouragement from time to time has been an inspiration for everyone.

The publisher and the editorial board hope that this book will prove to be a valuable piece of knowledge for students, practitioners and scholars across the globe.

Index

www.ingramcontent.com/pod-product-compliance
Lightning Source LLC
LaVergne TN
LVHW080715210926
848485LV00003B/470